3D Printing Applications in Cardiovascular Medicine

3D Printing Applications in Cardiovascular Medicine

Edited by

Subhi J. Al'Aref

Bobak Mosadegh

Simon Dunham

James K. Min

NewYork-Presbyterian Hospital and Weill Cornell Medicine

ACADEMIC PRESS

An imprint of Elsevier

Academic Press is an imprint of Elsevier
125 London Wall, London EC2Y 5AS, United Kingdom
525 B Street, Suite 1650, San Diego, CA 92101, United States
50 Hampshire Street, 5th Floor, Cambridge, MA 02139, United States
The Boulevard, Langford Lane, Kidlington, Oxford OX5 1GB, United Kingdom

Notices
Knowledge and best practice in this field are constantly changing. As new research and experience broaden our understanding, changes in research methods, professional practices, or medical treatment may become necessary.

Practitioners and researchers must always rely on their own experience and knowledge in evaluating and using any information, methods, compounds, or experiments described herein. In using such information or methods they should be mindful of their own safety and the safety of others, including parties for whom they have a professional responsibility.

To the fullest extent of the law, neither the Publisher nor the authors, contributors, or editors, assume any liability for any injury and/or damage to persons or property as a matter of products liability, negligence or otherwise, or from any use or operation of any methods, products, instructions, or ideas contained in the material herein.

Library of Congress Cataloging-in-Publication Data
A catalog record for this book is available from the Library of Congress

British Library Cataloguing-in-Publication Data
A catalogue record for this book is available from the British Library

ISBN: 978-0-12-803917-5

For information on all Academic Press publications visit our website at
https://www.elsevier.com/books-and-journals

Working together
to grow libraries in
developing countries

www.elsevier.com • www.bookaid.org

Publisher: John Fedor
Senior Acquisition Editor: Stacy Masucci
Editorial Project Manager: Sam W. Young
Production Project Manager: Poulouse Joseph
Designer: Victoria Pearson

Typeset by TNQ Technologies

Subhi J. Al'Aref, MD
I would like to dedicate this book to my parents, Jamal and Nisreen, as well as my siblings, Omar and Jumana, and my good friends, Gurpreet and Simran, for giving all their hearts and souls in order to provide me with the love, support, strength, and motivation to get to this stage of my career and contribute to this wonderful work. I would also like to thank Dr. Olaf S. Andersen for inspiring me to pursue a career as a physician scientist.

Bobak Mosadegh, PhD
I thank my family for all of their support throughout my career. I also would like to thank my previous supervisors, Prof. Noo Li Jeon, Prof. Shuichi Takayama, and Prof. George M. Whitesides, for their role in shaping me into a scientist and engineer capable of contributing to the academic community.

Simon Dunham, PhD
I would like to thank my family and friends for their support and motivation. Most of all, I would like to thank my wife, Dr. Annemarie Cardell, for her consistent patience and support throughout my training and career.

James K. Min, MD
My deepest appreciation goes to my parents and brother, and my best wishes to Nicholas and Alexander, whose generation I hope will realize and benefit from the vast potential of 3D printing in healthcare.

Contents

List of Contributors

Subhi J. Al'Aref
Department of Radiology, Weill Cornell Medicine, New York, NY, United States; Dalio Institute of Cardiovascular Imaging, NewYork-Presbyterian Hospital, New York, NY, United States

Ahmed Amro
Joan C. Edwards School of Medicine, Marshall University, Huntington, West Virginia, United States

Rami H. Awad
Technion Institute of Technology, Haifa

Jeroen J. Bax
Leiden University Medical Center, Leiden, The Netherlands

Muath Bishawi
Department of Biomedical Engineering, Pratt School of Engineering, Duke University, Durham, NC, United States; Division of Cardiothoracic Surgery, Department of Surgery, Duke University, Durham, NC, United States

Jonathan T. Butcher
Biomedical Engineering, Cornell University, Ithaca, NY, United States

Orlando Chirikian
Stanford Cardiovascular Institute, Stanford University School of Medicine, Stanford, California, United States

Simon Dunham
Department of Radiology, Weill Cornell Medicine, New York, NY, United States; Dalio Institute of Cardiovascular Imaging, NewYork-Presbyterian Hospital, New York, NY, United States

Mohamed B. Elshazly
Division of Cardiology, Department of Medicine, Weill Cornell Medicine-Qatar, Education City, Doha, Qatar

Brian P. Griffin
Cleveland Clinic, Cleveland, Ohio, United States

Sami A. Habash
University of Oxford, United Kingdom

Christopher J. Hansen
University of Massachusetts Lowell, Lowell, MA, United States

Serge C. Harb
Cleveland Clinic, Cleveland, Ohio, United States

Michael Hoosien
Department of Cardiovascular Medicine, Heart and Vascular Institute, Cleveland
Clinic, Cleveland, OH, United States

Daniel A. Hu
Stanford Cardiovascular Institute, Stanford University School of Medicine,
Stanford, California, United States

James B. Hu
Stanford Cardiovascular Institute, Stanford University School of Medicine,
Stanford, California, United States

Yasin Hussain
Department of Radiology, Weill Cornell Medicine, New York, NY, United States;
Dalio Institute of Cardiovascular Imaging, NewYork-Presbyterian Hospital, New
York, NY, United States

Amir Hossein Kaboodrangi
Department of Radiology, Weill Cornell Medicine, New York, NY, United States;
Dalio Institute of Cardiovascular Imaging, NewYork-Presbyterian Hospital,
New York, NY, United States

Kranthi K. Kolli
Department of Radiology, Weill Cornell Medicine, New York, NY, United States;
Dalio Institute of Cardiovascular Imaging, NewYork-Presbyterian Hospital, New
York, NY, United States

Thierry Le Jemtel
Tulane University School of Medicine, New Orleans, Louisiana, United States

Morteza Mahmoudi
Department of Anesthesiology, Brigham & Women's Hospital, Harvard Medical
School, Boston, Massachusetts, United States

Bobak Mosadegh
Department of Radiology, Weill Cornell Medicine, New York, NY, United States;
Dalio Institute of Cardiovascular Imaging, NewYork-Presbyterian Hospital,
New York, NY, United States

John Moscona
Tulane University School of Medicine, New Orleans, Louisiana, United States

Yazan Numan
Joan C. Edwards School of Medicine, Marshall University, Huntington,
West Virginia, United States

Hanley Ong
Department of Radiology, Weill Cornell Medicine, New York, NY, United States;
Dalio Institute of Cardiovascular Imaging, NewYork-Presbyterian Hospital,
New York, NY, United States

Sanlin Robinson
Sibley School of Mechanical & Aerospace Engineering, Cornell University, Ithaca, NY, United States

L. Leonardo Rodriguez
Cleveland Clinic, Cleveland, Ohio, United States

Eva A. Romito
Dalio Institute of Cardiovascular Imaging, NewYork-Presbyterian Hospital, New York, NY, United States

Qusai Saleh
Tulane University School of Medicine, New Orleans, Louisiana, United States

Vahid Serpooshan
Stanford Cardiovascular Institute, Stanford University School of Medicine, Stanford, California, United States; Department of Biomedical Engineering, Georgia Institute of Technology, Atlanta, Georgia, United States; Emory University School of Medicine, Atlanta, Georgia, United States

Robert Shepherd
Sibley School of Mechanical & Aerospace Engineering, Cornell University, Ithaca, NY, United States

Amanda Su
Department of Radiology, Weill Cornell Medicine, New York, NY, United States; Dalio Institute of Cardiovascular Imaging, NewYork-Presbyterian Hospital, New York, NY, United States

Kevin Luke Tsai
The Brooklyn Hospital Center, Brooklyn, New York, NY, United States

Alexander R. van Rosendael
Dalio Institute of Cardiovascular Imaging, NewYork-Presbyterian Hospital, New York, NY, United States

Sreekanth Vemulapalli
Division of Cardiology, Department of Medicine, Duke University, Durham, NC, United States

Sean M. Wu
Stanford Cardiovascular Institute, Stanford University School of Medicine, Stanford, California, United States; Department of Medicine, Division of Cardiovascular Medicine, Stanford University, Stanford, California, United States

Mohamed Zgaren
Dalio Institute of Cardiovascular Imaging, NewYork-Presbyterian Hospital, New York, NY, United States

Preface

Ever since I started learning CT angiography and understanding the powerful potential of modern medical imaging to create detailed 3D models of human vasculature, I have sought out methods to take advantage of this capability for diagnostic and therapeutic purposes. From here, my interest in 3D printing was a natural progression; these tools possess the power not only to make anatomically accurate models for visualization and testing but also to allow for the creation of patient-specific devices and implants by making designs from clinical images taken of the patient's anatomy, which are already available.

It was in this effort that I created a team of clinicians, computer scientists, and engineers at the Dalio Institute of Cardiovascular Imaging. Here, we aim to use the most advanced medical imaging coupled with advanced 3D printing methods to create truly patient-specific implants.

Working together as a team, our experience highlighted the technical challenges to achieving these types of implants and the critical need for a multidisciplinary approach to truly bring this technology to fruition. For the clinicians, it is critical to understand the capabilities and limitations of 3D printing materials and technology. For the engineers, it is essential to have a thorough and nuanced understanding of the underlying disease pathology in order to invent the needed solutions. It is only after working as a collaborative team that we had the ability to move past the conventional ways of manufacturing devices with generic shapes, sizes, and materials.

The experience of assembling a multidisciplinary team is what motivated me to have a book published that covered both the technology and the clinical needs of 3D printing in cardiovascular medicine. It is my hope that this book inspires engineers, scientists, and clinicians, to come together and collaborate to develop the next generation of 3D printed cardiac innovations. Despite the many hurdles that are discussed in this book, the future of this field is bright, and I look forward to the next wave of innovation that will inevitably follow this text.

Dr. James K. Min

Acknowledgments

This textbook is based on the recent applications of 3D printing for the evaluation and treatment of cardiovascular ailments. We would like to acknowledge the contribution of every author in making this textbook a reality, as well as the support from the publisher. We would like to express our sincere gratitude to the diligent team at the Dalio Institute of Cardiovascular Imaging, including Miss Niree Hindoyan, Mrs. Jittaporn "June" Thumpituk, and Dr. Millie Gomez, for their dedication, support, and contributions to the content herein presented. We would also like to acknowledge the generous support of Mr. Ray Dalio of the Dalio Institute of Cardiovascular Imaging, without whom none of this would have been possible.

Introduction

Cardiovascular surgery methods have advanced substantially since 1896, when Dr. Ludwig Rehn of Germany performed the first successful cardiac surgery by ligation of a stab wound to the heart accessed through a thoracotomy. Since the 1990s, open heart surgery has been performed using cardiopulmonary bypass that provides oxygen to the body while the heart is steadied and emptied of its blood, allowing for more complicated and protracted surgeries to be performed. At the same time, many surgeons were being trained on newer minimally invasive surgical techniques, which typically rely on laparoscopic and endoscopic tools to manipulate tissue via numerous small incisions; this is in contrast to open heart surgery that requires a large enough access route to fit the surgeons' hands. When noninvasive imaging techniques became clinically established and used on a widespread scale, minimally invasive cardiac procedures, focusing on the heart or surrounding vascular structures, became an active area of research and clinical interest. To date, this technique has primarily been used for treating coronary artery disease, but recently a myriad of transcatheter techniques have been developed to address the various forms of structural heart disease. These advances in cardiovascular surgery and intervention have been driven by engineering, coupled with unheralded advances in imaging and surgical techniques, which established the platform to provide more efficacious and safer treatment.

3D printing (i.e., fabrication of an object using layer-by-layer deposition) is a technology that has been around since the 1980s, and has made impacts on a variety of industries by enabling cost-effective prototyping of complex geometric objects. This technology is primarily used for prototyping of visual models due to the limited availability of printable materials, which also have weaker material properties than those processed with traditional manufacturing methods. As 3D printing matures as a technology, however, it will likely be used ubiquitously to manufacture objects on-demand since it allows for creation of intricate shapes and interfaces not possible with other methods. Furthermore, it is the most effective method for creating custom-designed objects since arbitrary shapes can be created without investment in building molds.

For the field of cardiovascular medicine, 3D printing has been primarily utilized to create anatomic models and tools for procedural planning. The reason for this initial use is for several reasons: (1) medical imaging of the heart is regularly performed, (2) procedural planning often requires assessment of 3D geometries, thus providing a useful application of this technology, and (3) these models have been shown to be useful for the training of surgeons and interventionalists, as well as being informative for the patients and their families. Another use of 3D printing is to make patient-specific models for in vitro setups, allowing assessment of hemodynamics and performance of cardiac devices. 3D printing has yet to be used to create permanent endocardial/vascular implants due to the unique challenges regarding hemocompatibility and durability of the printed materials, although academic research has begun to create such patient-specific implants. This book will delve into the current state of 3D printing for cardiovascular medicine, explaining its advantages and disadvantages for particular indications and procedures, as well as the future developments needed to further utilize this technology for this field.

History of 3D Printing

Amanda Su[1,2]**, Subhi J. Al'Aref**[1,2]

Department of Radiology, Weill Cornell Medicine, New York, NY, United States[1]*; Dalio Institute of*
Cardiovascular Imaging, NewYork-Presbyterian Hospital, New York, NY, United States[2]

INTRODUCTION TO 3D PRINTING

Traditionally, industries used subtractive manufacturing to construct products, whereby designs were carved out of a solid block of material. In contrast, additive manufacturing uses a layer-by-layer technique that allows for more intricate designs and interior modeling. Rapid prototyping was one of the first uses of additive manufacturing. It allowed manufacturers to create prototypes much faster, facilitating the evaluation and testing of designs before producing a finished product. In rapid prototyping, 3D models are first created using computer-aided design software. Machines then construct 3D objects based on that model.

EARLY RESEARCH

The earliest research into the use of photopolymers to create 3D objects took place in the 1960s at Battelle Memorial Institute in Ohio. The aim of the experiment was to polymerize resin by intersecting two laser beams of differing wavelengths. Wyn Swainson applied for a patent in 1971 for a similar dual laser beam approach called photochemical machining [1]. He subsequently founded the Formigraphic Engine Company in California, but this technology never materialized into a commercially available system [2]. In the late 1970s, Dynell Electronics Corporation invented solid photography. This technology used a laser or milling machine to cut cross-sections based on a computer model, and then stack them together to form an object [3].

BEGINNING OF 3D PRINTING

Hideo Kodama, at the Nagoya Municipal Industrial Research Institute in Japan, was one of the first to develop a rapid prototyping technique using a single laser beam [4]. Though he submitted a patent application for this invention in 1980, it expired without proceeding to the later stages of the Japanese patent process. In 1980 and

3D Printing Applications in Cardiovascular Medicine. https://doi.org/10.1016/B978-0-12-803917-5.00001-8

1981, he published papers on his experiments to develop methods for automatic fabrication of three-dimensional models using UV rays and a photosensitive resin, using a mask to control exposure of UV source. He described techniques of solidifying thin consecutive layers of photopolymer [5], key aspects of what would later be called stereolithography (SLA) (Fig. 1.1).

In 1984, Charles Hull invented stereolithography. He was issued a patent for stereolithography in 1986, and in the patent described a process in which liquid polymers were hardened under UV light to form cross-sections of a 3D model [6]. This method used digital data and a computer-controlled beam of light to create each layer, one on top of the other. Hull subsequently founded 3D Systems, which eventually produced and sold stereolithography machinery. The first commercial SLA printer in the world was produced by 3D Systems in 1988 [2,3].

Around the same time as Hull's SLA patent, Carl Deckard, at the time still an undergraduate student at the University of Texas, developed the concept of the selective laser sintering (SLS) process. SLS was based on the selective solidification of powder using a laser beam [7]. Deckard went on to found Desktop Manufacturing Corporation (DTM Corp), which produced its first SLS printers in 1992. DTM was eventually acquired by 3D Systems. In 1993, Deckard founded Sinterstation 2000, which launched SLS technology into the industry [8].

S. Scott and Lisa Crump founded the company Stratasys, and in 1989 filed a patent for a form of rapid prototyping called fused deposition modeling (FDM), in which a plastic filament or metal wire was heated in a nozzle and extruded. Its deposition was guided by a computer, based on a predetermined digital model. Each layer was kept at a temperature just below solidification point for good inter-layer adhesion [9]. Stratasys eventually developed thermoplastic and printer systems for 3D printing [3].

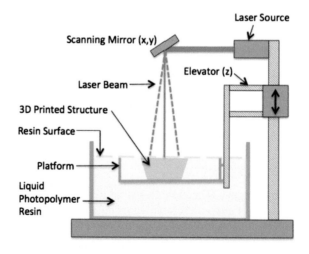

FIGURE 1.1

Diagram illustrating the components of a stereolithography system.

Later in 1989, Hans Langer in Germany formed Electro Optical Systems (EOS), with a focus on direct metal laser sintering, which fabricated 3D parts directly from computer design models. This technology used selective exposure of a laser to metal powder for liquid phase sintering [10]. EOS sold its first stereo system in 1994, and is recognized today for industrial prototyping. EOS acquired the right to all DTM patents related to laser sintering in 2004 [11].

In the early 1990s, several other 3D printing techniques were being investigated. Ballistic Particle Manufacturing, patented by William Masters, projected micro-droplets of molten wax material from a jet moving in an X−Y plane to form thin cross-sections. The stationary platform moved in the Z-axis to allow for each layer of the 3D object to be added [12]. Michael Feygin filed a patent for laminated object manufacturing in 1995, which used automated formation of cross-sectional slices from sheet material according to a digital 3D model, then stacking and bonding the layers to form a solid object. However, Feygin's company, Helisys Inc., soon went out of business due to financial difficulties [13]. Solid ground curing was invented by Itzchak Pomerantz, and used an optical mask system to selectively expose layers of photocurable resin. The remaining liquid was then removed and replaced with wax, which was then milled to form a flat substrate for the next layer [14].

In the mid-1990s, the 3D printing industry split into 2 areas of focus: high end for highly engineered complex parts (e.g., medical) and printers for concept development and functional prototyping—user-friendly, cost effective. By the end of the 1990s, only three original companies remained: 3D Systems, Statasys, and EOS [3].

COMMERCIALIZATION OF 3D PRINTING

The patent for Statasys' FDM technology expired around 2005, and subsequently two open source 3D printer projects started: the RepRap Movement and Fab@ Home. Both projects had the goal of developing and sharing designs for a 3D printer that was affordable to a wider range of individuals.

Adrian Bowyer, a senior lecturer in mechanical engineering at the University of Bath, started an open source project to create a 3D printer based on FDM technology that became known as the RepRap movement. Bowyer developed designs for a self-replicating rapid-prototyper, and published his designs online and encouraged others to post their improved versions [15,16].

The first printer, Darwin, was released in 2007 and the second, named Mandel, in 2009 (Fig. 1.2). Individuals could download these files to make the plastic parts for the printer, and necessary metal parts were ideally available at any hardware store. However, these early printers were difficult to put together and did not always function well. Josef Prusa from the Czech Republic released his own design, the Prusa Mendel, in 2010, and this simplified version accelerated the development of better printer models. Eventually, the goal became to create whole printer kits that would be easy to use for consumers. The first available kit based on the RepRap

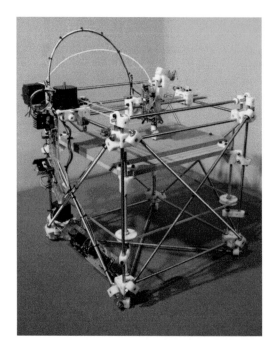

FIGURE 1.2

Original Darwin printer.

Image courtesy of reprap.org, reproduced with permission.

concept was the BfB RapMan, which was released in January 2009. This was closely followed by the MakerBot Cupcake CNC in April 2009 and the MakerBot Thing-O-Matic in 2010 [16,17].

Fab@Home was developed by Evan Malone, a PhD candidate at Cornell University at the time and Dr. Hod Lipson, who was an assistant professor of mechanical and aerospace engineering at Cornell University [18]. Its simple, low-cost design was based on solid free-form fabrication, with a multiple syringe-based deposition method that allowed for printing of multiple materials at once [19]. In order to foster growth in the user community, they created an online website with detailed instructions and downloadable software, as well as an online user forum to facilitate discussions between users. In the first 6 months, the website had over 4.3 million requests for pages from more than 150,000 hosts [18]. People were able to use this technology in a wide variety of applications, from printing chocolate and cheese to batteries and conductive wires. Notably, one project, Fab@School, was developed to allow use of 3D printers in elementary schools, and could print with materials such as play-doh [20]. The Fab@Home project closed in 2012, as 3D printers became widely available to consumers and the project's goal had been achieved.

Crowdfunding sites like Kickstarter facilitated the launching of new 3D printing companies by allowing entrepreneurs to gain funding from public support. In 2012, with funding from Kickstarter, the B9Creator, based on DLP technology, was introduced to the market in June, closely followed by the Form 1 stereolithography printer that raised nearly $3 million. In 2013, the Buccaneer, a filament-based printer, raised about $1.5 million on Kickstarter. Since then, many related projects printing various objects, including post-processing technologies have been successful on this platform [16,21].

3D PRINTING IN MEDICINE

There have been many developments in the healthcare field based on 3D printing. The first application of 3D printing in healthcare involved printing of patient-specific, anatomically accurate structures for surgery and orthopedics, such as surgical planning [22] and custom implants [23]. In 2005, it was shown that 3D printed hydroxyapatite scaffolds could be designed based on anatomical information of a particular patient from acquired radiologic images. At the same time, the internal structure of bone could be replicated and used to facilitate live cell growth in preparation for implantation [24]. Since then, various groups have successfully 3D printed other body parts including skull, jaw, hip, and sacrum [25,26] (Fig. 1.3).

In 2003, Dr. Thomas Boland at Clemson University filed the first patent for a technique that involved printing of viable cells. Most human cells are 25 microns in diameter, and could fit through a typical inkjet nozzle. Dr. Boland attached an HP printer to a platform that could move along the Z-axis, allowing for 3D printing of cells [27]. In 2004, Gabor Forgacs filed a patent for a scaffold-free bioprinting technique that could print multiple cells at the same time. Instead of using an inkjet's nozzle to print one cell at a time, Dr. Forgacs used a syringe with a needle or

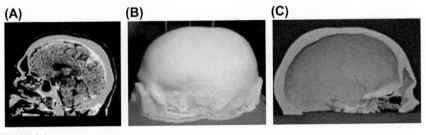

FIGURE 1.3

Patient's head CT used to construct skull rendering and configure 3D surface for printing. (A) CT scan visualized using InVesalius. The green highlights the skull to be rendered, selected using image intensity. (B) 3D printed skull with structural support from lateral viewpoint. (C) 3D printed skull from medial viewpoint after removing the internal support material.

Reproduced with permission from Naftulin JS, et al. Streamlined, inexpensive 3D printing of the brain and skull. PLoS One 2015;10:e0136198.

FIGURE 1.4

Organovos syringe-based extrusion printer used to produce biofacial tissues.

Reproduced from http://organovo.com/about/news/media-library/ with permission.

micropipette that could fit many more cells. In this extrusion-based process, pressure was applied to push the cell aggregates out in spheroid or cylindrical form. Additionally, because scaffolds can interfere with a cell's natural biology, this approach relied on cells' intrinsic ability to develop structure, and used only a gel to hold the cells together until they began to grow and fuse together [28]. The company Organovo was founded in 2007, with the intent of continuing Forgacs' technology, and is one of the first bioprinting companies. In 2009, Organovo was awarded the first NIH grant for bioprinting blood vessels [29]. Its syringe-based extrusion printer which has been particularly successful in producing biofacial tissues, used mostly in research and development. (Fig. 1.4) Also in 2009, Dr. Anthony Atala's group at Wake forest printed the first fully functional skin constructs. In 2011, Dr. Atala discussed the challenges and potentials for 3D bioprinting in organ regeneration, and as proof of concept printed live cells onto a kidney-like scaffold. Though it was a nonfunctioning model, it showcased much of the progress that has been made in this field [30].

3D printing has also been widely used in biomedical research. Microfluidic cell culture devices have been created using 3D printing technologies to replicate various organ systems on microchips. With living cells arranged to mimic physiologically accurate microenvironments, new drugs and therapies can be tested and developed ex vivo [31]. In 2013, Organovo created the first bioprinted fully cellular liver tissue, now being used in research to model liver disease [32].

CHARLES HULL

Charles (Chuck) Hull was born in May 12, 1939, in Clifton, Colorado, and received a BS in engineering physics from the University of Colorado in 1961. He started his career as a senior engineer at Bell and Howell, and then spent 10 years as an engineering manager at DuPont's Photo Products division, working primarily on the development of analytical equipment for chemists such as mass spectrometers. It was as vice president of engineering at UVP, Inc. in California that he developed and patented the SLA process. His work at the time involved using UV light to harden tabletop coatings, and this inspired his use of UV light in the SLA process. The first object he built successfully using this technology was a small cup 5 cm tall, first seen by his wife in the middle of the night. In 1984, he cofounded 3D systems, which quickly became popular in the automobile and aerospace industries, and was widely used for medical applications as well [33,34] (Fig. 1.5).

In 1996, Mr. Hull was awarded the William T Ennor Manufacturing Technology Award by American Society of Mechanical Engineers. Also in 1996, he was named Entrepreneur of the Year for high technology by the Ernst & Young Institute. In 1997, he won the Albert M Sargent Progress Awards by Society of Manufacturing Engineers. In 2014, he received the European Inventor Award, and was inducted into the National Inventors Hall of Fame.

Mr. Hull is currently the cofounder, executive vice president, and chief technology officer of 3D Systems. In 2014, 3D Systems launched the Ekocycle Cube printer in collaboration with singer Will.i.am. The Ekocycle Cube printer uses recycled materials and is meant for everyday use, priced at $1199 [35]. 3D Systems brought in $633 million in 2016, and has three bases in the United Kingdom [36]. Future projects include collaborating with Google on the next generation phone technology.

3D printing was invented less than 50 years ago, and has been utilized in mainstream industry for less than a decade. In this relatively short time, the use of 3D printing has expanded into a multitude of areas, and developed from large-scale manufacturing to being accessible for personal use. In particular, 3D printing in healthcare has advanced at a rapid pace, facilitating more precise and

FIGURE 1.5

Charles W. Hull.

Reproduced with permission from the European Patent Office.

individualized interventions, and now playing an integral role in cutting-edge tissue engineering research. With its versatility and continued development, 3D printing is likely to have a great influence on healthcare in the future.

REFERENCES

[1] Patent US4041476-Method, medium and apparatus for producing three-dimensional figure product—Google patents [Internet]. Available from: https://www.google.com/patents/US4041476.

[2] Beaman JJ, Barlow JW, Bourell DL, Crawford RH, Marcus HL, McAlea KP. Solid freeform fabrication: a new direction in manufacturing: with research and applications in thermal laser processing [Internet]. Illustrated. Springer Science & Business Media; 2013. Available from: https://books.google.com/books?id=IPvxBwAAQBAJ&pg=PA47&lpg=PA47&dq=swainson+patent&source=bl&ots=xxIkDFBwUD&sig=I_ErMVsAyundvQ07g3-BaPt3yJs&hl=en&sa=X&ved=0ahUKEwiMp8rL76bUAhVH_IMKHbngAXIQ6AEIUDAM#v=onepage&q=swainson%20patent&f=false.

[3] Wohlers T, Gornet T. History of additive manufacturing. Wohlers Associates, Inc.; 2014.

[4] Beaman JJ, Barlow JW, Bourell DL, Crawford RH, Marcus HL, McAlea KP. Solid freeform fabrication: a new direction in manufacturing. Boston, MA: Springer US; 1997.

[5] Kodama H. Automatic method for fabricating a three-dimensional plastic model with photo-hardening polymer. Rev Sci Instrum 1981;52(11):1770−3.

[6] Patent US4575330-Apparatus for production of three-dimensional objects by stereolithography—Google patents [Internet]. Available from: https://www.google.com/patents/us4575330.

[7] Patent US5597589-Apparatus for producing parts by selective sintering—Google patents [Internet]. Available from: https://www.google.com/patents/US5597589.

[8] Carl Deckard selected for AMUG innovators award | additive manufacturing (AM) [Internet]. Available from: http://additivemanufacturing.com/2016/11/03/carl-deckard-selected-for-amug-innovators-award/.

[9] Patent US5340433-Modeling apparatus for three-dimensional objects—Google patents [Internet]. Available from: https://www.google.com/patents/US5340433.

[10] Khaing MW, Fuh JYH, Lu L. Direct metal laser sintering for rapid tooling: processing and characterisation of EOS parts. J Mater Process Technol 2001;113(1−3):269−72.

[11] Santos EC, Shiomi M, Osakada K, Laoui T. Rapid manufacturing of metal components by laser forming. Int J Mach Tool Manufact 2006;46(12−13):1459−68.

[12] Cooper K. Rapid prototyping technology: selection and application [Internet]. CRC Press; 2001. Available from: https://books.google.com/books?id=H50qMZFN4JEC&pg=PA21&lpg=PA21&dq=ballistic+particle+manufacturing&source=bl&ots=Lsi4M0v9g6&sig=NVdNdK6AEIqVkqXpV3rfRWegIQ8&hl=en&sa=X&ved=0ahUKEwjlzsus5rPTAhVL8IMKHeEBCQAQ6AEIYjAN#v=onepage&q=ballistic%20particle%20manufacturing&f=false.

[13] Chua CK, Leong KF, Lim CS. Rapid prototyping: principles and applications (with companion CD-ROM). WORLD SCIENTIFIC; 2010.

[14] Patent US5031120-three dimensional modelling apparatus—Google patents [Internet]. Available from: https://www.google.com/patents/US5031120.

[15] Micallef J. Beginning design for 3D printing [Internet]. Apress; 2015. Available from: https://books.google.com/books?id=_YfDCgAAQBAJ&pg=PA22&lpg=PA22&dq= reprap+movement&source=bl&ots=MFt0m_BrFP&sig=lGmdkxDU0-jdo0ApzfTK QFEwhA0&hl=en&sa=X&ved=0ahUKEwjPrZfdoLbTAhWF1IMKHXR3DoU4Ch DoAQgsMAI#v=onepage&q=reprap%20movement&f=false.

[16] Horvath J. Mastering 3D printing (technology in action) [Internet]. 1st ed. Apress; 2014 Available from: https://www.amazon.com/Mastering-3D-Printing-Technology-Action/ dp/1484200268.

[17] The Official History of the RepRap Project. All about 3D printing. 2016. Available from: https://all3dp.com/history-of-the-reprap-project/.

[18] Malone E, Lipson H. Fab@Home: the personal desktop fabricator kit. Rapid Prototyp J 2007;13(4):245−55.

[19] Lipson H, Kurman M. Fabricated: the world of 3D printing. Indianapolis: John Wiley & Sons, Inc.; 2013.

[20] Steele B. Low-cost, home-built 3-D printer could launch a revolution, say Cornell engineers | Cornell Chronicle. Cornel Chronicle 2007. Available from: http://news. cornell.edu/stories/2007/02/low-cost-home-built-3-d-printer-could-launch-revolution.

[21] The free beginner's guide—history—3D printing industry [Internet]. Available from: https://3dprintingindustry.com/3d-printing-basics-free-beginners-guide/history/.

[22] Robiony M, Salvo I, Costa F, et al. Virtual reality surgical planning for maxillofacial distraction osteogenesis: the role of reverse engineering rapid prototyping and cooperative work. J Oral Maxillofac Surg 2007;65(6):1198−208.

[23] Mazzoli A. Selective laser sintering in biomedical engineering. Med Biol Eng Comput 2013;51(3):245−56.

[24] Leukers B, Gülkan H, Irsen SH, et al. Hydroxyapatite scaffolds for bone tissue engineering made by 3D printing. J Mater Sci Mater Med 2005;16(12):1121−4.

[25] Whitaker M. The history of 3D printing in healthcare. Bull Roy Coll Surg Engl 2014; 96(7):228−9.

[26] Kim D, Lim JY, Shim KW, et al. Sacral reconstruction with a 3D-printed implant after Hemisacrectomy in a patient with sacral osteosarcoma: 1-year follow-up result. Yonsei Med J 2017;58(2):453−7.

[27] Patent US20040237822-Ink-jet printing of viable cells—Google patents [Internet]. Available from: https://www.google.com/patents/US20040237822?dq=ininventor:% 22Thomas+Boland%22+bioprinting&hl=en&sa=X&ved=0ahUKEwjp7oW0s67UA hXKbz4KHUBZClcQ6AEIMDAC.

[28] Patent US8241905-Self-assembling cell aggregates and methods of making engineered tissue … —Google Patents [Internet]. Available from: https://www.google.com/patents/ US8241905?dq=ininventor:%22forgacs%22+bioprinting&hl=en&sa=X&ved=0ahU KEwjt0u_3s67UAhVB2D4KHUa5Dvg4FBDoAQghMAA.

[29] History—organovo [Internet]. Available from: http://organovo.com/about/history/.

[30] Harmon K. A sweet solution for replacing organs. Sci Am 2013;308(4):54−5.

[31] Bhatia SN, Ingber DE. Microfluidic organs-on-chips. Nat Biotechnol 2014;32(8): 760−72.

[32] Nguyen DG, Funk J, Robbins JB, et al. Bioprinted 3D primary liver tissues allow assessment of organ-level response to clinical drug induced toxicity in vitro. PLoS One 2016;11(7):e0158674.

[33] Chuck Hull: the father of 3D printing who shaped technology | Business | the Guardian [Internet]. Available from: https://www.theguardian.com/business/2014/jun/22/chuck-hull-father-3d-printing-shaped-technology.

[34] Charles W. Hull Co-Founder and Chief Technology Officer. Available from: https://en.wikipedia.org/.

[35] 3D printing, recycling and Will.i.am unite in the Ekocycle Cube | Guardian Sustainable business | the Guardian [Internet]. Available from: https://www.theguardian.com/sustainable-business/2014/sep/18/cola-cola-3d-printing-filabot-reprap-plastic-recycling-black-eyed-peas.

[36] 3D systems reports fourth quarter and full year 2016 financial results | 3D Systems [Internet]. Available from: https://www.3dsystems.com/press-releases/3d-systems-reports-fourth-quarter-and-full-year-2016-financial-results.

3D Printing Methods

2

Rami H. Awad[1], Sami A. Habash[2], Christopher J. Hansen[3]

Technion Institute of Technology, Haifa[1]; University of Oxford, United Kingdom[2]; University of Massachusetts Lowell, Lowell, MA, United States[3]

INTRODUCTION
OVERVIEW OF 3D PRINTING PROCESS

Three-dimensional (3D) printing is the process of translating a three-dimensional digital model into a physical instantiation (Fig. 2.1). This translation process uses a 3D printer, that is, a machine that renders the digital geometry as a spatially defined assembly of material(s). The translation of a precisely defined digital object into a physical copy necessarily entails a geometric approximation—such as rounded corners, surface roughness, trapped voids, or other imperfections. These approximations are imposed both by the machine, by the material components assembled, and the trade-offs involved in size, resolution, and production rates. Parameters that are critical to the cardiovascular specialist are outlined in this introduction, and will be referenced throughout each method.

The chapter will discuss the entire 3D printing process from the digital part file to the 3D printer and the printed pattern. The digital model and file are primarily independent of the print method, and the slicing software is tailored to the machines, so the digital translation into 3D printer instructions will be discussed separately from the printing methods. The remainder of the chapter is devoted to discussing

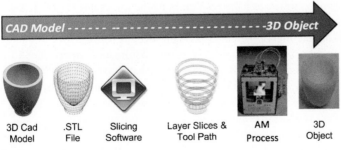

FIGURE 2.1

General process flow for 3D print methods.

Reproduced with permission from Campbell T, et al. Could 3D printing change the world. In: Technologies, potential, and implications of additive manufacturing. Washington, DC: Atlantic Council; 2011.

3D Printing Applications in Cardiovascular Medicine. https://doi.org/10.1016/B978-0-12-803917-5.00002-X

five primary types of 3D printing methods that dominate both academic research and industrial fabrication. These five methods are:

1. Fused deposition modeling (FDM)
2. Selective laser sintering (SLS)
3. Direct write (DW)
4. Stereolithography (SLA)
5. Material jetting (MJ)

For each method, the strengths and limitations of the printing technique will be discussed so as to guide the designer to the most capable and productive method for their cardiovascular application.

PROCESSING PARAMETERS RELEVANT TO CARDIOVASCULAR PRINTING

Current 3D printing methods are presently being investigated for structural heart disease applications and pertaining to valvular diseases, pediatric and adult congenital heart disease, coronary arteries and systemic vasculature [1]. Such applications broadly include [2–4]:

1. Advanced visualization and diagnosis.
2. Planning and simulating surgery and interventions.
3. Education, communication, and research.

Though more applications in cardiovascular medicine are presented throughout the book, the aforementioned bullets are highlighted to offer a general reference in relation to the selection of 3D printing methodologies.

In compliance with relevant regulations, technical requirements, and practical concerns [5,6], 3D printers used for cardiovascular applications ought to be chosen with consideration to the following criteria:

1. *Resolution*: A printed resolution hinges on a combination of the resolution of the machine motion and the resolution of the minimum material "voxel" (i.e., volumetric pixel) size that can be added to the assembly. The smaller the resolution, the higher the correspondent level of detail.
   For Cartesian-axis systems (i.e., $X-Y-Z$ coordinate systems), the horizontal resolution is the smallest distance the printer can move in the 2D plane defined by the $X-Y$ axes, while the vertical resolution is the minimal layer thickness the printer can produce via motion in the Z-axis direction. Non-Cartesian robotic arm-based systems operate with rotational coordinate systems as well, but have similar resolution considerations in the equivalent 3D space. Building a general model depicting a relatively large organ, say a model of the heart often employed in medical education demonstrations, would tolerate a reduced resolution relative to those demanded by surgeons for a pre-interventional perception of complex anatomy.

2. *Size*: The size of a specific product is limited to that of the build platform volume, comprising the base area X and Y dimensions and the height Z dimension, or comprising the radial extension of robotic arms. For each of the technologies discussed in this chapter, the build size of most professional printer desktops and academic research machines offers a sufficient volume to accommodate any cardiovascular product of choice.

3. *Materials*: A cardiovascular designer likely has a specific material or set of materials selected for printing. The initial material format (e.g., solid filament, powder, or liquid) and ability to process the material by itself or in a multimaterial modality heavily influences the choice of a printing technique. The processing approach also affects final material properties and functional performance, such as mechanical resistance to static or dynamic loads, durability in biological environments, and surface roughness. Aspects pertaining to materials' selection criteria are thoroughly discussed in Chapter 3.

4. *Finishing*: 3D printed objects often require some postprocessing. Geometries with overhangs that require support material must have the support material removed after printing. The material may require a postcure under a UV light or in an oven to fully develop the desired properties. Other processing, such as painting, surface smoothening, or cleaning, requires additional equipment and time.

5. *Time*: The time required to complete a given fabrication run is a function of the selected method and the resolution. Some methods pattern material in a serial fashion, such that the build time and build volume are roughly linearly correlated; by contrast, printing methods based on parallel patterning may not slow down within a layer, but could be dominated by the overhead of preparing to pattern the next layer. A finer resolution requires not only more 2D layers to be patterned, but often also a greater precision of the machine that may entail a slower patterning speed per layer. Some machines can provide estimated build times, such that the interplay of the aforementioned effects can be considered prior to fabrication.

6. *Training*: The end user should be trained to master the substages of 3D printing relevant to their use of the machine. This may include managing the software, preprocessing the model, loading the material, operating the machinery, and postprocessing the product. The initial design—consisting of the geometry, material, and process—requires professionally knowledgeable engineers with sufficient capability in mechanical and manufacturing fundamentals.

7. *Cost*: Costs of 3D printing can be considered using an external manufacturer (i.e., a contractor or authorized third party agent) or an inhouse fablab. Anecdotes abound of companies and labs with a printer gathering dust in a corner due to the lack of institutional resources to maintain an inhouse 3D printing operation; so, a realistic assessment of the trade-off between inhouse or outsourced fabrication should be considered. 3D printing as-a-service [7] can be performed without any upfront investment except to create the part file. For commercial machines and materials, a platform service (e.g., 3D Hubs) can provide an instant quote for cost

and production time; a more custom design or material usually requires a specialized contractor. For cutting-edge research and products, an inhouse fablab enables maximal control over design iterations. The initial cost would comprise the cumulative price of the software package(s) to be used, a lab setup, and any related health and safety upgrades, and the printer device itself. Operating costs include the material input, maintenance, waste disposal, and staff. These and other costs are detailed in reference [8].

DIGITAL MODELS

Each 3D printed object begins as a digital model. This digital file is commonly called a computer-aided design (CAD) model (Fig. 2.1). The CAD model may be generated by an end user exploiting a 3D modeling software, scanned based on an existing object (e.g., computed tomography, magnetic resonance imaging, optical scanner) [3,9], or purchased from a service provider. Regardless of the origin of the file, users can perform subsequent modifications during an iterative design process using a CAD software package.

CAD software is available both in free, open-source and professional versions. A vibrant community of hobbyists and other low-cost users benefit from free, open-source CAD software systems. Professional CAD packages, which may possess finite element simulation functionalities, can require thousands of US dollars for a license. Though each CAD platform may store the geometry in a separate (often proprietary) file format, most are able to convert the file into the standardized stereolithography (STL) file format consisting of a triangular mesh representing the geometry surface in 3D Cartesian space [10].

The 3D surface geometry defined by the STL file is then converted using a "slicing" software into 2D layers. For a serial patterning method, each layer then has an algorithmically generated toolpath that defines the coordinates to be printed and the order in which they will be patterned. Because the coordinates only define the geometry of the outer surface, the user can define the fill pattern used to fill the part interior; the fill can range from fully filled (i.e., no void space) to sparsely filled (e.g., only 10% filled with 90% void space) [11,12]. If the 3D geometry has layers that are larger than the layers below, support material (i.e., a sacrificial material region that can be removed during postprocessing) may be required to support the layer from collapsing under gravity [13]. The final toolpath(s) consists of the object surface geometry, its fill pattern, and the necessary support material.

These toolpaths are converted into a numerical control programming language such as G-code, which controls the motion of the printing stage. While open-source research and hobbyist platforms still possess clear demarcations between these processing steps, commercial printers commonly perform these steps automatically within their proprietary software packages. Finally, the G-code file is sent to the 3D printer to command the motion of the stage. Any commands for material flow or exposure of the material to a phase-change stimulus (e.g., UV light) are also issued coordinated with the motion. The resulting combination of motion and material assembly is a physical approximation of the original 3D digital object.

PRINTING METHODS

The number and variety of 3D printing methods have expanded significantly over the previous decades. In this section, 5 of the most common printing methods are introduced. The processes are presented in order of the physical mechanism by which the printing operates, such that the most similar printing methods are discussed in succession. Specifically, the section begins with solid materials that undergo melting and resolidification (FDM) or sintering (SLS), followed by viscoelastic materials that possess both solid and liquid characteristics (DW), and finally liquid-based techniques using a bath (SLA) or jetted liquid droplets (MJ, PJ). The ASTM Committee on Additive Manufacturing has defined additional broad categories of printing [14]—binder jetting, directed energy deposition, and sheet lamination—that are not covered here due to their more limited application to the cardiovasculature; please refer to reference [15] for an introduction to these techniques.

FUSED DEPOSITION MODELING

FDM or fused filament fabrication is one of the most widespread types of desktop 3D printers [16]. FDM is a well-established and relatively user-friendly print method that accepts a wide variety of materials, including multiple-material printing. The machines are relatively compact, modestly priced, and require minimal laboratory modifications. Due to these advantages, FDM has become a de facto standard for prototyping and the production of custom parts, such as fixtures and holders, as well as an entry level machine for 3D printing experimentation for biologically relevant materials [17–19].

The FDM process is an extrusion-based process and is illustrated in Fig. 2.2. The process relies on a solid filament that is actively pulled into the extruder head by a drive gear; the extruder head contains a heated nozzle assembly that locally melts the material. The feed of solid material pushes the molten material to extrude through the nozzle orifice into the build volume and quickly resolidifies upon cooling. The extruder head is mounted to a motion-controlled stage, whose motions are coordinated with the drive gear to extrude material at a rate matched to the velocity of the overall assembly. The technique is capable of multimaterial printing using multiple extrusion heads. Often a second head is included for printing of a support material, which is required to support material overhangs beyond an approximately 45 degrees angle.

The strength of FDM is the broad range of materials that can be printed; if the material can be spooled and melted by heat, it can often be printed [20]. Thermoplastic materials dominate FDM, though specialty formulations and new

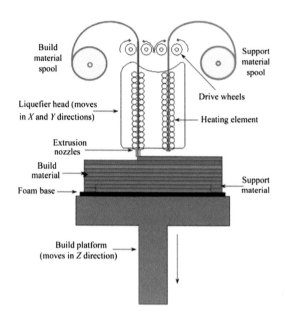

FIGURE 2.2

Schematic of the fused deposition modeling 3D printing method.

Reproduced with permission from Mohamed OA, et al. Optimization of fused deposition modeling process parameters: a review of current research and future prospects. Adv Manuf 2015;3:42–53.

materials are the subject of ongoing research [21–23]. The filament is typically extruded and spooled in one of two standard sizes, 1.75 and 3.0 mm in diameter. Other diameters are possible, particularly in research settings and on open-source machines; however, the solid filament diameter must be sufficiently large to be gripped by the drive gear and to effectively seal the molten material within the extruder head so that it can pressurize the material and drive it through the orifice. The filament uncoiling is usually facilitated by means of a passive mechanism.

Thermal considerations dominate the FDM process [24]. The temperature of the nozzle should be set to fully melt the material within the assembly, as insufficient melting will lead to inconsistent prints. However, if the temperature is set too high, the material will heat far beyond its melting point, leading to reduced viscosity and a lengthier time to resolidify, both effects which will reduce the feature consistency and fidelity. The heat transfer between the nozzle and the entering material will also depend on the print velocity, where higher print speeds correspond to a reduced residence time in the nozzle and a shorter time for heating [25]. Hence, for machines in which the nozzle temperature can be set, the user should consider changes to the nozzle set temperature to account for any changes made to the material and print speed.

Resolution in FDM is generally proportional to the diameter of the nozzle orifice and the precision of the axis mechanical movement [20]. The combination of these factors results in a range from 500 microns in a basic amateur FDM desktop printer to as fine as 25 microns in optimized professional or academic printers. The smallest feature width that can be printed is roughly the diameter of the nozzle orifice, though under- or overpumping can result in smaller or larger (up to 2×) features than the orifice diameter. In practice, a single filament feature may be too mechanically weak to be retained in the structure during printing and postprocessing steps. The smallest layer thickness, likewise, corresponds to the nozzle diameter. FDM parts possess a visible surface step structure that makes fine details unachievable, without postprocessing. The typical orifice diameter is 200–500 microns, but orifices 50 microns or smaller have been demonstrated [26].

The build volume varies over two orders of magnitude, depending on the printer. A typical range is from $100 \times 100 \times 100$ mm^3 for a standard desktop device to $1000 \times 600 \times 1000$ mm^3 for standalone machines; yet larger printers—such as the Big Area Additive Manufacturing machine at Oak Ridge National Labs—can print a $6 \times 2.25 \times 1.8$ m^3 build volume [27]. Hence, a printer with sufficient build area to print any conceivable cardiovascular models at full scale is available. Recently, researchers have been exploring robotic arms and other stages with greater than 3 axes of motion for deposition of material onto parts or supports with complex curvatures [28].

The print time depends on the volume of the digital model, its internal fill, the feature complexity, its orientation during printing, and the volumetric rate of material deposition. A typical extruder head traverses the build volume at a speed of 50–150 mm/s. While this traversal speed can be increased, there is typically an unacceptable resolution trade-off for the existing hardware, while new motion-controlled hardware would entail greater costs to retain the desired resolution. If the user chooses a smaller feature size, the number of layers and the number of deposition passes per layer will increase, and the print time will be proportionally increased.

Postprocessing is required for parts that have support material, for surfaces that require a smoother finish, or to meet dimensional constraints. Support material removal is performed either mechanically (through physical breaking, grinding, etc.) or by means of dissolving in solution. Options to achieve a smoother surface include hand sanding, detergents, vapor treatments [29], or postmachining operations [30]. For parts that must meet dimensional constraints, particularly for mating assemblies, the interlocking part geometries typically require sanding, filing, or other material removal from rounded corners or rough surfaces.

Costs associated with FDM are relatively low compared to other print methods. Research labs, which often favor open-source equipment, may obtain a satisfactory machine for 1000 USD. So called "prosumer" machines range from 5000 to 20,000 USD, while high-end professional production scale systems with large build volumes can cost 50,000–200,000 USD. The material costs range from 25 USD per kilogram for standard thermoplastics (Acrylonitrile Butadiene Styrene (ABS) or Polylactic Acid (PLA)) to significantly higher for specialized products containing reinforcements, catalysts, and other novel chemistries.

SELECTIVE LASER SINTERING

SLS printers are based on the principle of laser sintering of a powdered material [31]. SLS is a subcategory of "powder bed fusion," in which a powdered material is fused by some directed energy source. In sintering, the material is heated, without reaching its melting point, up to sufficiently high temperatures promoting the fusion of the powder into a solid mass. The process may accommodate complex geometries—including interior components or channels—from a wide variety of materials (polymers, metals, ceramics) with high part densities, strengths, and stiffness; it can also incorporate multicomponent materials (e.g., alloys) in a single powder [32,33].

The SLS process is illustrated in Fig. 2.3. The powder to be sintered is located in a heated build bed that has been smoothed to a flat surface by a traversing roller. The path the laser will traverse is imported from the sliced STL file, and is often optimized by the machine for uniform temperature distribution and reduced build times [34]. The laser beam is directed by lenses and mirrors to pattern its directed energy onto the powdered material resulting in a fusion of the exposed powder into a solid block. At the conclusion of the layer, the bed is lowered by one layer thickness, and the supply bed rises to provide a source of powder for the next layer. A recoating roller passes the source powder onto the newly exposed surface and levels the powder surface before the next layer is patterned. Hence, this method is a layer-by-layer fabrication technique. The unsintered supply powder that surrounds the sintered object acts as the support material, and so no surface artifacts from support material are present.

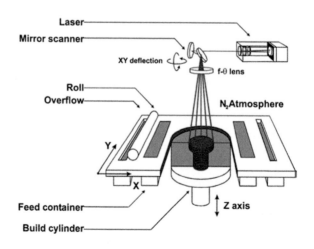

FIGURE 2.3

Schematic of selective laser sintering 3D printing method.

Reproduced with permission from Kruth JP, et al. Binding mechanisms in selective laser sintering and selective laser melting. Rapid Prototyping J 2005;11:26—36.

The resolution of an SLS printer is dependent on the powdered material, the laser optics and power, and the machine parameters [33,35]. In general, SLS offers a medium-scale resolution when compared to other printing technologies, with a standard resolution as high as 50−100 microns. The resolution in the $X−Y$ plane is dominated by the laser optics, the resulting spot size, and the thermal conductivity of the material. The Z-direction resolution, by contrast, is dictated by the chosen layer thickness, which typically ranges from 50 to 200 microns. The total build volume is dictated by the powder bed size, and ranges from $200 \times 250 \times 300$ mm^3 on the smaller end to $700 \times 400 \times 600$ mm^3 at the larger end. Because the bed must be fully filled to the depth of the printed object, a larger bed will require substantially more powder; hence, a machine should be chosen to match the object size(s) expected for the targeted application. Substantial research and commercial effort has recently been expended in recovery of unsintered powder following a print [36], and commercial vendors often sell a separate powder recovery stage for customers who wish to recover powder. Unsintered powder may have been exposed to an elevated temperature that causes drift in the material properties [37], so providers have guidelines for users to avoid adversely affecting the resolution of subsequent prints using recovered powder.

The time associated with building a part depends on the laser raster speed [38], the powder transfer process, layer thickness, choice of geometry size, and orientation [39,40]. SLS machines have dramatically improved build speeds in recent years [41]—a full production volume of 700 cm^3/h is possible—and is working to shed a perception of being a slow printing technology. The $X−Y$ speed depends on the laser raster over the object pattern; this process is relatively fast, as the laser speeds are typically in the order of 2−6 m/s. By contrast, the Z-direction changes are slow, requiring the physical process of raising new material and rolling it into a smooth layer for the next print pattern. Hence, the slow Z-direction changes dominate the printing process, and changes that reduce the number of layers can have an outsize impact on print times. If the resolution of fine layers (e.g., 50 microns) can be exchanged for that of thicker layers (e.g., 200 microns), the build time will be substantially reduced. The shortest axial dimension of the part to be printed should be oriented in the Z-direction so as to minimize the total number of layers required.

Postprocessing of the part does not require the removal of any support, only the removal of excess powder loosely bound to the surface. Often a soft-bristle brush or a pressurized air stream is sufficient for this task. However, the part surfaces are relatively rough and so users frequently subject them to postprinting polishing [42,43] or a chemical treatment [44]. The surface layer can be slightly porous, so if a nonporous surface is required, a thin polymer layer is often applied; as a by-product, this polymer layer often smoothens the surface and can reduce the need for a polishing step. A postbake is often prescribed to improve mechanical properties [45].

The cost and practical considerations associated with SLS printing are higher than those associated with FDM. The machine capital cost is medium to high

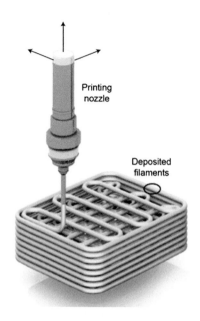

FIGURE 2.4

Schematic of direct write 3D printing method.

Reproduced with permission from A.R. Studart, Additive manufacturing of biologically-inspired materials, Chem Soc Rev 45, 2016,359–376.

(100,000+ USD), while the powder material cost is relatively moderate (50–150 USD per kg). The operation of SLS printing machines requires adherence to best practices for safe operation of high power laser systems.

DIRECT WRITING

DW or direct write assembly is an extrusion-based approach [46] that goes by various names, including gel printing, paste printing, micro-extrusion, direct ink writing, and robocasting. Similar to the FDM method, a continuous filament is extruded through a nozzle orifice; by contrast with FDM, however, the patterned material is a viscoelastic material—that is— a material with either primarily elastic (solidlike) or viscous (liquidlike) behavior under quiescent or shear conditions, respectively. Because the material will flow under shear forces, a thermally induced phase change (as in FDM) is unnecessary. The materials palette can therefore include biological materials that must remain under physiological conditions, and is well suited to the range of materials with elasticity values corresponding to biological tissues [47]. The technique has been used to print biocompatible materials [48,49], cells [50], and synthetic microvasculature [51,52].

The process of DW is illustrated in Fig. 2.4. The viscoelastic material is loaded within a syringe or other reservoir and fitted with a nozzle, typically a commercially available dispense tip. The reservoir is mounted to a motion-controlled Cartesian stage or robotic arm system to produce layer-by-layer or free-form fabrication, respectively. The ink reservoir can be pressurized either by a displacement-controlled or a pressure-controlled piston, and the ink is pressurized in coordination with the stage motion. Upon sufficient pressure to exceed the ink shear yield stress, the material flows through the nozzle orifice as a continuous filament onto the substrate; upon cessation of shear stresses, the patterned material regains the elastic component of its viscoelastic modulus and becomes functionally solid [53]. Multimaterial printing can be achieved using single [54] or multiple extrusion heads, and support materials can be patterned for large span distances.

The resolution of DW, similarly, to FDM, is a combination of effects from the motion stage resolution and the orifice diameter of the nozzle. While larger stages (an axial direction >500 mm) may have screw-driven axes with limited precision, the stage resolution rarely dominates the effects of the nozzle orifice at small or intermediate stage sizes (i.e., smaller than 200 mm axial direction). The feature diameter is roughly equal to that of the nozzle orifice, though under- or overpumping can achieve diameters 75%−300% the diameter of the orifice [51]. These feature diameters control the surface roughness of the part, and lead to a stairstep surface similar to that seen in FDM. Because the material is viscoelastic, it has the ability to deform and flow under gravitational forces over long time periods [55]; hence, the material stiffness should be designed to minimize this effect over the expected lifetime of the part. The build volume for small, highly precise stages (resolution <0.1 micron) is roughly $100 \times 100 \times 20 \text{ mm}^3$ or smaller. For stages with lower precision (resolution >5 microns), build volumes of $1000 \times 1000 \times 200 \text{ mm}^3$ or larger are possible.

Fabrication timescales depend on the nozzle translation velocity, the feature size, and the timescale associated with the viscoelastic modulus recovery in the ink material. Nozzle translation velocities are typically 10−100 mm/s, though print speeds up to 1 m/s have been demonstrated. The build time will increase with the square of the inverse of the nozzle diameter for a fully filled part, as the number of filaments per layer and the number of layers both proportionally increase. The viscoelastic modulus of the ink is multiple orders of magnitude lower than that of the plateau modulus for the zero-shear state [56]; upon cessation of shear stresses, some materials require tens of seconds to hours to fully recover the plateau modulus. For these materials, a significantly lower translation rate is required to produce features with acceptable fidelity. Materials containing biological cells may also require slow print speeds in order to minimize mechanical damage to the living cell. Though typically a serial processing method, a high throughput parallelized nozzle has been investigated for single- and multimaterial deposition that reduces print times by 2+ orders of magnitude [57,58].

Postprocessing of the structures is minimal, as further mechanical forces will deform the viscoelastic material structure. Photopolymerizable inks can be exposed to UV light, while colloidal-based systems may be sintered within a heated oven.

Synthetic microvasculature is often embedded within a liquid reactive resin, thereby rendering an inverse pattern of the printed structure [59—61]. In these instances, the ink is meant to be a fugitive ink that can be removed after a thermally induced phase transformation [62] or by solvent dissolution [60].

The cost of DW is relatively moderate compared to other methods, and commensurate with those of FDM. Suitable motion-stages can be purchased for as little as 5—10K USD, while ancillary dispensing systems are available for 1—2K USD. Material costs highly depend on the material system, but can range from 5 USD per kilogram for a wax-based ink to in excess of 1K USD per kilogram for specialty functional inks. The operation of DW printing machines requires more knowledge of the ink rheology and print parameter interactions, and is less automated than other 3D printing methods.

STEREOLITHOGRAPHY

SLA printing is the first 3D printing technology (see Chapter 1). The process is based on a directed energy beam that photopolymerizes a liquid resin [63]. Results may reflect among the highest degree of resolution and complexity achievable in 3D printing of millimeter-scale objects. Dominant use cases in medical modeling and prototyping include the creation of 3D models for diagnosis, lightweight concept models, preoperative planning, and custom-fitting implant design. The technique requires the use of photopolymers, of which formulations may be susceptible to mechanical degradation over time.

The SLA process is illustrated in Fig. 2.5. The original approach to SLA contains a resin bath that is filled with a liquid photopolymer that becomes reactive

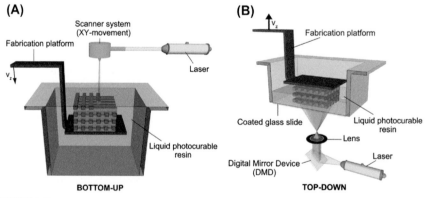

FIGURE 2.5

Schematic of (A) Traditional stereolithography and (B) Continuous liquid interface production 3D printing methods.

Reproduced with permission from Billiet T, et al. A review of trends and limitations in hydrogel-rapid prototyping for tissue engineering. Biomaterials 2012;33:6020—41.

when exposed to light of a sufficiently small wavelength (i.e., sufficient energy) and polymerizes to form a solid. A UV laser or other light source is directed by means of lenses and mirrors to raster over the surface of a fluid resin bath, thereby locally polymerizing the resin to form a solid 2D layer. The build platform is lowered into the bath so that the solidified layer is at a depth equal to one layer thickness. A wiper traverses the surface to smooth the resin into a uniform plane, and the process is ready to pattern the next layer. Recently, a more rapid and cost-effective CLIP (continuous liquid interface production) approach has been developed [64]. The CLIP method has an optical window at the bottom of the resin bath, which permits the resin to be photopolymerized by light exposure from below. The polymerized layer is prevented from adhering to the window by the diffusion of oxygen through the window, which inhibits the reaction within an approximately 20−30 micron layer [65]. The part is then built from the topmost layer downward, and is physically pulled layer-by-layer vertically from the bath, as opposed to being lowered into the bath layer-by-layer in SLA. The advantages to the CLIP approach are multiple: the print is more rapid, as no physical wiper needs to create the next surface layer, and a smaller amount of liquid resin is required for printing, as only a thin layer covering the window is necessary rather than a bath of resin as deep as the part. Another advance, digital light processing (DLP), uses a digital micromirror device (DMD) common in optical projection systems to simultaneously expose the resin to a light pattern [66,67]. This exposure turns the serial process of a laser raster into the parallel process of a pattern exposure. As such, the DLP approach can enhance print speeds, though it is limited to the resolution of the image projected by the DMD array.

Resolution in SLA is influenced by the precision of the mirrors directing the laser, the optical features of the light source, and the properties of the photopolymer. An $X-Y$ resolution of 75 microns is typical, while a Z resolution of 25 microns is regularly achieved. The design of the photopolymer system is critical to the patterned voxel size; the reader is referred to Chapter 3 and to reference [68] for a detailed review of the formulation effects on resolution. The total build volume is dictated either by the bath volume (in traditional SLA) or the window area by the travel height of the base plate for the CLIP-based approach. Typical SLA builds can be $1500 \times 750 \times 550$ mm^3, while CLIP builds have been commercially demonstrated to be $190 \times 120 \times 325$ mm^3. The size of CLIP-based machines are not yet physically limited, and larger build sizes will likely be realized in future machine generations that match or exceed that of SLA-based systems.

The time associated with fabrication depends on the feature and layer resolution and complexity [69,70]. Due to laser tracing speed or DLP simultaneous projection and fast polymerization, SLA is considered to possess among the fastest build rates of all 3D printing methods. Moreover, the CLIP process has further sped the process by removing the need for a physical wiper motion, leading to a "continuous" production motion, though the projected patterns are still layer-by-layer slices. If a lower resolution can be tolerated, larger laser spot sizes can cure features in order of magnitude larger in the horizontal build plane and show a concomitant reduction in build times.

Postprocessing for SLA consists of support material removal, any surface touch-up [71], and a postcure step. Supports, typically in the form of pillars, are required in the bath-based process to prevent disconnected overhang features from floating away [72]. Supports provide this function in the CLIP-based approach, as well as support for the weight of the part as it is drawn from the liquid. Chemical baths are used to remove any excess resin adhering to the surface and to smoothen the surface [73]. A postcure of "green" object (i.e., object after initial print) is often performed to react any residual reactive groups and to enhance the mechanical properties and long-term durability. This postcure depends on the chemistry and is often performed either under strong ultraviolet light or within an oven set to a moderate temperature (e.g., $<120°C$) [74].

With respect to costs and practical considerations, SLA machines constitute a larger capital outlay than FDM, but affordable desktop devices exist. Likewise, photopolymer resins are relatively expensive—up to thousands of USD per gallon—and have a limited shelf life. Resin life can be extended by storage in a cool location, shielded from light exposure. The operation of SLA machines is straightforward and requires only basic user preparation.

MATERIAL JETTING

MJ is a process in which liquid droplets of material are selectively deposited onto a build bed to fabricate a 3D structure [75]. Photopolymer jetting (PJ) is a subset of MJ, in which the droplets are photopolymerizable. The broader MJ materials palette includes non-photopolymer materials, such as low melting point waxes, while PJ and its liquid photopolymer materials dominate this general process. The technique is known for its high resolution and intricate features, smooth surface finish, multimaterial and multicolor printing, and diversity of mechanical properties ranging from rigid to flexible and tough objects [76]. However, as with SLA, the mechanical properties of the printed object may degrade over the long term [77]. The features of materials palette and resolution are attractive in replicating biological tissues in which fine features, coloration, and component stiffness are key attributes for medical applications. Castings with intricate details are a common use of wax-based printed objects, and the dental industry is a large market segment for multiple commercial vendors [78].

The MJ/PJ process is illustrated in Fig. 2.6. MJ printers use two kinds of input material: a build material, which constitutes the finished object, and a support material with a lower melting temperature. PJ printers can use two or more input materials: one or more build materials and the support material. For MJ by thermal melting of a build material, the material is heated to melt the material and then jetted

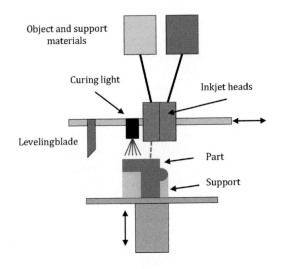

FIGURE 2.6

Schematic for material jetting 3D printing method.

Reproduced with permission from Stansbury JW, Idacavage MJ, et al. 3D printing with polymers: challenges among expanding options and opportunities. Dent Mater 2016;32:54–64.

onto a build platform, where upon cooling it solidifies to form the object. For MJ of a photopolymer material (i.e., the PJ process), the printhead jets droplets of the photopolymer, which are subsequently UV-cured to solidify the layer. In each case, the droplet jetting is coordinated with the $X-Y$ stage motion. For photocurable materials, the deposited material is then exposed to UV light mounted to the $X-Y$ axes of motion. The build platform is then lowered by a height equal to the layer thickness to prepare for the deposition of a subsequent layer. Because the material is deposited as a liquid, any overhangs must be built atop a support material, which is deposited using a dedicated printhead.

MJ competes with SLA for the highest resolution 3D printing technology. The resolution in MJ is dictated by the inkjet printhead resolution, the droplet physical properties, and for photopolymers, the curing dynamics [79,80]. Inkjet printheads are similar to those of commercial paper inkjet printheads, and achieve a similar maximum of approximately 600 dots per inch. The droplet physical properties— such as the viscosity and surface tension—dictate the spreading of the droplet upon contact with the build bed or previously deposited material [81]. While a molten droplet will quickly cool upon deposition, a photopolymer will remain liquid until exposed to light and therefore has greater opportunity for spreading and loss of resolution. Some professional systems have a roller pass over the liquid photopolymer layer prior to UV curing as it achieves a more uniform Z-height that increases the resolution of subsequent layers. For optimized professional printer and material systems, resolutions of 40 microns in each of the $X-Y$ directions and 15 microns in the Z direction are achievable [82]. The total build volume can be as large as $520 \times 380 \times 300$ mm^3.

While MJ excels in resolution, the production rate is among the slowest of all 3D print methods. The build time depends on the object geometry and orientation, selected resolution, and the number of materials incorporated [83]. To achieve high $X-Y$ resolutions, the droplet layer thickness is typically only ~ 15 microns thick. While this high Z-resolution is advantageous for certain applications, the extreme number of layers results in slow vertical builds. As such, the operator is advised to place the shortest axial dimension in the Z-direction to drastically reduce build time; an exception to this rule is if such an orientation would require extra support material, which also increases build time.

Postprocessing of parts is relatively minimal. If support structures were required, they are either mechanically removed, melted away (for MJ), or easily washed away by means of a solution bath (for PJ). The surface quality is typically sufficiently fine to obviate the need for polishing, though the region in contact with support material will have a more matte finish that contrasts with the reflective surface of the rest of the part. Most PJ materials are recommended to undergo a postcure under a UV lamp or a postbake in a moderately heated oven.

The costs associated with MJ or PJ printing are among the highest of all printing approaches. The capital cost of the machines have been high (100K USD or greater), their production speed low, and material costs higher per mass ($\sim 150-400$ USD per kg) than any other print method [84]. Though operation is relatively easy, the development of new materials is sufficiently complex so as to require specialized development machines for proper optimization of droplet size under inkjet ejection, viscosity, surface tension, and other formulation parameters.

SUMMARY

The five 3D print methods reviewed here are those with greatest relevance to cardiovascular and biological printing. The preparation of the digital printing file by converting a CAD file to an STL and slicing it for the specific machine and process was discussed. An overview of the broad physical behavior during printing and the factors relevant to the designer or practitioner were outlined, including: suitable material types, the resolution of the finest features, the build rate and volume, machine and material cost, and the resulting functional properties. These factors are qualitatively summarized in Table 2.1. The trade-offs between these performance metrics and the suitability for the specific material(s) of interest for 3D printing are critical considerations that should be examined in the specification of any new 3D printing research or production program.

Table 2.1 Qualitative Comparison of 3-D Print Methods Relevant to Cardiovascular Printing

Performance Metric	Fused Deposition Modeling	Selective Laser Sintering	Direct Write	Stereolithography/Continuous Liquid Interface Production	Material Jetting/ Photopolymer Jetting
Materials	1–2+	1	1–4+	1	1–7+
Support	Often	Rarely	Occasionally	Often	Often
Resolution	Low	High	Low-high	High	High
Build rate	Low	High	Low-medium	Very high	Low-medium
Machine cost	Low-medium	Medium-high	Low-medium	Low-high	Medium-high
Material cost	Low-medium	Low	Low-high	Medium-high	High-very high
Mechanical properties	Low-high	High	Very low-low	Medium	Medium-high

REFERENCES

[1] Giannopoulos A, Mitsouras D, Yoo S-J, Liu PP, Chatzizisis YS, Rybicki FJ. Applications of 3D printing in cardiovascular diseases. Nat Rev Cardiol 2016;13:701.

[2] Horn TJ, Harrysson OL. Overview of current additive manufacturing technologies and selected applications. Sci Prog 2012;95:255−82.

[3] Marro A, Bandukwala T, Mak W. Three-dimensional printing and medical imaging: a review of the methods and applications. Curr Probl Diagn Radiol 2016;45:2−9.

[4] Tack P, Victor J, Gemmel P, Annemans L. 3D-printing techniques in a medical setting: a systematic literature review. Biomed Eng Online 2016;15:115.

[5] Patra S, Young V. A review of 3D printing techniques and the future in biofabrication of bioprinted tissue. Cell Biochem Biophys 2016;74:93−8.

[6] Morrison RJ, Kashlan KN, Flanangan CL, Wright JK, Green GE, Hollister SJ, et al. Regulatory considerations in the design and manufacturing of implantable 3D-printed medical devices. Clin Transl Sci 2015;8:594−600.

[7] Baumann FW, Roller D. Additive manufacturing, cloud-based 3D printing and associated services—overview. J Manuf Mater Process 2017;1:15.

[8] Thomas DS, Gilbert SW. Costs and cost effectiveness of additive manufacturing. NIST - Spec Publ 2014;1176:12.

[9] Rengier F, Mehndiratta A, Von Tengg-Kobligk H, Zechmann CM, Unterhinninghofen R, Kauczor H-U, et al. 3D printing based on imaging data: review of medical applications. Int J Comput Assist Radiol Surg 2010;5:335−41.

[10] Baumann FW. 3D printing-as-a-service for collaborative engineering. 2017.

[11] Wu J, Aage N, Westermann R, Sigmund O. Infill optimization for additive manufacturing—approaching bone-like porous structures. IEEE Trans Visual Comput Graph 2018;24:1127−40.

[12] Campagna F, Diaz AR. Optimization of lattice infill distribution in additive manufacturing. In: ASME 2017 International design engineering technical conferences and computers and information in engineering conference; 2017. p. V02AT03A 028−V02AT03A028.

[13] Strano G, Hao L, Everson R, Evans K. A new approach to the design and optimisation of support structures in additive manufacturing. Int J Adv Manuf Technol 2013;66: 1247−54.

[14] Monzón M, Ortega Z, Martínez A, Ortega F. Standardization in additive manufacturing: activities carried out by international organizations and projects. Int J Adv Manuf Technol 2015;76:1111−21.

[15] Srinivas M, Babu BS. A critical review on recent research methodologies in additive manufacturing. Mater Today: Proc 2017;4:9049−59.

[16] Mohamed OA, Masood SH, Bhowmik JL. Optimization of fused deposition modeling process parameters: a review of current research and future prospects. Adv Manuf 2015;3:42−53.

[17] Long J, Gholizadeh H, Lu J, Bunt C, Seyfoddin A. Application of fused deposition modelling (FDM) method of 3D printing in drug delivery. Curr Pharmaceut Des 2017;23:433−9.

[18] Kalita SJ, Bose S, Hosick HL, Bandyopadhyay A. Development of controlled porosity polymer-ceramic composite scaffolds via fused deposition modeling. Mater Sci Eng C 2003;23:611−20.

[19] Cao T, Ho K-H, Teoh S-H. Scaffold design and in vitro study of osteochondral coculture in a three-dimensional porous polycaprolactone scaffold fabricated by fused deposition modeling. Tissue Eng 2003;9:103–12.

[20] Turner BN, Gold SA. A review of melt extrusion additive manufacturing processes: II. Materials, dimensional accuracy, and surface roughness. Rapid Prototyping J 2015;21:250–61.

[21] Tian X, Liu T, Yang C, Wang Q, Li D. Interface and performance of 3D printed continuous carbon fiber reinforced PLA composites. Compos Appl Sci Manuf 2016;88:198–205.

[22] Brenken B, Favaloro A, Barocio E, Denardo N, Kunc V, Pipes RB. Fused deposition modeling of fiber-reinforced thermoplastic polymers: past progress and future needs. In: Proceedings of the American Society for Composites: thirty-first technical conference; 2016.

[23] Gkartzou E, Koumoulos EP, Charitidis CA. Production and 3D printing processing of bio-based thermoplastic filament. Manuf Rev 2017;4:1.

[24] Bellini A, Guceri S, Bertoldi M. Liquefier dynamics in fused deposition. J Manuf Sci Eng 2004;126:237–46.

[25] Yardimci MA, Hattori T, Guceri SI, Danforth S. Thermal analysis of fused deposition. In: Proceedings of solid freeform fabrication conference; 1997. p. 689–98.

[26] Monzón MD, Gibson I, Benítez AN, Lorenzo L, Hernández PM, Marrero MD. Process and material behavior modeling for a new design of micro-additive fused deposition. Int J Adv Manuf Technol 2013;67:2717–26.

[27] Duty CE, Kunc V, Compton B, Post B, Erdman D, Smith R, et al. Structure and mechanical behavior of big area additive manufacturing (BAAM) materials. Rapid Prototyping J 2017;23:181–9.

[28] Wulle F, Coupek D, Schäffner F, Verl A, Oberhofer F, Maier T. Workpiece and machine design in additive manufacturing for multi-axis fused deposition modeling. Procedia CIRP 2017;60:229–34.

[29] Espalin D, Medina F, Arcaute K, Zinniel B, Hoppe T, Wicker R. Effects of vapor smoothing on ABS part dimensions. In: Proceedings from rapid 2009 conference & exposition, Schaumburg, IL; 2009.

[30] Pandey PM, Reddy NV, Dhande SG. Improvement of surface finish by staircase machining in fused deposition modeling. J Mater Process Technol 2003;132:323–31.

[31] Gibson I, Shi D. Material properties and fabrication parameters in selective laser sintering process. Rapid Prototyping J 1997;3:129–36.

[32] Olakanmi EOT, Cochrane R, Dalgarno K. A review on selective laser sintering/melting (SLS/SLM) of aluminium alloy powders: processing, microstructure, and properties. Prog Mater Sci 2015;74:401–77.

[33] Shirazi SFS, Gharehkhani S, Mehrali M, Yarmand H, Metselaar HSC, Kadri NA, et al. A review on powder-based additive manufacturing for tissue engineering: selective laser sintering and inkjet 3D printing. Sci Technol Adv Mater 2015;16:033502.

[34] Senthilkumaran K, Pandey PM, Rao P. Influence of building strategies on the accuracy of parts in selective laser sintering. Mater Des 2009;30:2946–54.

[35] Yadroitsev I, Shishkovsky I, Bertrand P, Smurov I. Manufacturing of fine-structured 3D porous filter elements by selective laser melting. Appl Surf Sci 2009;255:5523–7.

[36] Faludi J, Baumers M, Maskery I, Hague R. Environmental impacts of selective laser melting: do printer, powder, or power dominate? J Ind Ecol 2017;21.

[37] Slotwinski JA, Garboczi EJ, Stutzman PE, Ferraris CF, Watson SS, Peltz MA. "Characterization of metal powders used for additive manufacturing. J Res Natl Inst Stand Technol 2014;119:460.

[38] Lexow MM, Drexler M, Drummer D. Fundamental investigation of part properties at accelerated beam speeds in the selective laser sintering process. Rapid Prototyping J 2017;23:1099–106.

[39] Narayana B, Reddy LM, Vekatesh S. Selective laser sintering: effect of part orientation on building time. 2017.

[40] Zhang Y, Bernard A. Generic build time estimation model for parts produced by SLS. In: High value manufacturing: advanced research in virtual and rapid prototyping. Proceedings of the 6th international conference on advanced research in virtual and rapid prototyping; 2013. p. 43–8.

[41] Shen F, Yuan S, Chua CK, Zhou K. Development of process efficiency maps for selective laser sintering of polymeric composite powders: modeling and experimental testing. J Mater Process Technol 2018;254:52–9.

[42] King BAT, Rennie AEW, Taylor JP, Bennett GR. Post processing treatments on laser sintered nylon 12. 2017.

[43] Guo J, Bai J, Liu K, Wei J. "Surface quality improvement of selective laser sintered polyamide 12 by precision grinding and magnetic field-assisted finishing. Mater Des 2018;138:39–45.

[44] Crane N, Ni Q, Ellis A, Hopkinson N. Impact of chemical finishing on laser-sintered nylon 12 materials. Addit Manuf 2017;13:149–55.

[45] Taylor SS. Thermal history correlation with mechanical properties for polymer selective laser sintering (SLS). 2017.

[46] Hon K, Li L, Hutchings I. Direct writing technology—advances and developments. CIRP Annals 2008;57:601–20.

[47] Xie B, Parkhill RL, Warren WL, Smay JE. Direct writing of three-dimensional polymer scaffolds using colloidal gels. Adv Funct Mater 2006;16:1685–93.

[48] Sun L, Parker ST, Syoji D, Wang X, Lewis JA, Kaplan DL. Direct-write assembly of 3D silk/hydroxyapatite scaffolds for bone co-cultures. Adv Healthcare Mater 2012;1: 729–35.

[49] Fu Q, Saiz E, Tomsia AP. Direct ink writing of highly porous and strong glass scaffolds for load-bearing bone defects repair and regeneration. Acta Biomater 2011;7:3547–54.

[50] Bertassoni LE, Cardoso JC, Manoharan V, Cristino AL, Bhise NS, Araujo WA, et al. Direct-write bioprinting of cell-laden methacrylated gelatin hydrogels. Biofabrication 2014;6:024105.

[51] Wu W, DeConinck A, Lewis JA. Omnidirectional printing of 3D microvascular networks. Adv Mater 2011;23.

[52] Wu W, Hansen CJ, Aragón AM, Geubelle PH, White SR, Lewis JA. Direct-write assembly of biomimetic microvascular networks for efficient fluid transport. Soft Matter 2010;6:739–42.

[53] Therriault D, White SR, Lewis JA. Rheological behavior of fugitive organic inks for direct-write assembly. Appl Rheol 2007;17:10112–20.

[54] Hardin JO, Ober TJ, Valentine AD, Lewis JA. Microfluidic printheads for multimaterial 3D printing of viscoelastic inks. Adv Mater 2015;27:3279–84.

[55] Smay JE, Cesarano J, Lewis JA. Colloidal inks for directed assembly of 3-D periodic structures. Langmuir 2002;18:5429–37.

[56] Conrad JC, Ferreira SR, Yoshikawa J, Shepherd RF, Ahn BY, Lewis JA. Designing colloidal suspensions for directed materials assembly. Curr Opin Colloid Interface Sci 2011;16:71—9.

[57] Hansen CJ, Saksena R, Kolesky DB, Vericella JJ, Kranz SJ, Muldowney GP, et al. High-throughput printing via microvascular multinozzle arrays. Adv Mater 2013;25:96—102.

[58] Kranz S. Multinozzle printheads for 3D printing of viscoelastic inks. 2013.

[59] Therriault D, Shepherd RF, White SR, Lewis JA. Fugitive inks for direct-write assembly of three-dimensional microvascular networks. Adv Mater 2005;17:395—9.

[60] Hansen CJ, Wu W, Toohey KS, Sottos NR, White SR, Lewis JA. Self-Healing materials with interpenetrating microvascular networks. Adv Mater 2009;21:4143—7.

[61] Toohey KS, Hansen CJ, Lewis JA, White SR, Sottos NR. Delivery of two-part self-healing chemistry via microvascular networks. Adv Funct Mater 2009;19:1399—405.

[62] Therriault D, White SR, Lewis JA. Chaotic mixing in three-dimensional microvascular networks fabricated by direct-write assembly. Nat Mater April 2003;2:265—71.

[63] Melchels FP, Feijen J, Grijpma DW. A review on stereolithography and its applications in biomedical engineering. Biomaterials 2010;31:6121—30.

[64] Tumbleston JR, Shirvanyants D, Ermoshkin N, Janusziewicz R, Johnson AR, Kelly D, et al. Continuous liquid interface production of 3D objects. Science 2015;347:1349—52.

[65] Janusziewicz R, Tumbleston JR, Quintanilla AL, Mecham SJ, DeSimone JM. Layerless fabrication with continuous liquid interface production. Proc Natl Acad Sci U S A 2016; 113:11703—8.

[66] Sun C, Fang N, Wu D, Zhang X. Projection micro-stereolithography using digital micro-mirror dynamic mask. Sens Actuator Phys 2005;121:113—20.

[67] Zheng X, Lee H, Weisgraber TH, Shusteff M, DeOtte J, Duoss EB, et al. Ultralight, ultrastiff mechanical metamaterials. Science 2014;344:1373—7.

[68] Halloran JW. Ceramic stereolithography: additive manufacturing for ceramics by photopolymerization. Annu Rev Mater Res 2016;46:19—40.

[69] Chen CC, Sullivan PA. Predicting total build-time and the resultant cure depth of the 3D stereolithography process. Rapid Prototyping J 1996;2:27—40.

[70] Campbell I, Combrinck J, de Beer D, Barnard L. Stereolithography build time estimation based on volumetric calculations. Rapid Prototyping J 2008;14:271—9.

[71] Robert EW, Vicki LM. Abrasive flow finishing of stereolithography prototypes. Rapid Prototyping J 1998;4:56—67.

[72] Kazemi M, Rahimi A. Supports effect on tensile strength of the stereolithography parts. Rapid Prototyping J 2015;21:79—88.

[73] Cunico MWM, Cunico MM, Cavalheiro PM, de Carvalho J. Investigation of additive manufacturing surface smoothing process. Rapid Prototyping J 2017;23:201—8.

[74] Watters MP, Bernhardt ML. Curing parameters to improve the mechanical properties of stereolithographic printed specimens. Rapid Prototyping J 2017;24(1):46—51.

[75] Gibson I, Rosen D, Stucker B. Material jetting. In: Additive manufacturing technologies: 3D printing, rapid prototyping, and direct digital manufacturing. New York, NY: Springer New York; 2015. p. 175—203.

[76] Vaezi M, Chianrabutra S, Mellor B, Yang S. Multiple material additive manufacturing—part 1: a review: this review paper covers a decade of research on multiple material additive manufacturing technologies which can produce complex geometry parts with different materials. Virtual Phys Prototyping 2013;8:19—50.

[77] Costa CA, Linzmaier PR, Pasquali FM. Rapid prototyping material degradation: a study of mechanical properties. IFAC Proc Vol 2013;46:350—5.

[78] Alharbi N, Wismeijer D, Osman RB. Additive manufacturing techniques in prosthodontics: where do we currently stand? A critical review. Int J Prosthod 2017;30.

[79] Kumar K, Kumar GS. An experimental and theoretical investigation of surface roughness of poly-jet printed parts: this paper explains how local surface orientation affects surface roughness in a poly-jet process. Virtual Phys Prototyping 2015;10: 23–34.

[80] Yap YL, Wang C, Sing SL, Dikshit V, Yeong WY, Wei J. Material jetting additive manufacturing: an experimental study using designed metrological benchmarks. Precis Eng 2017;50:275–85.

[81] Derby B. Inkjet printing of functional and structural materials: fluid property requirements, feature stability, and resolution. Annu Rev Mater Res 2010;40:395–414.

[82] Yang H, Lim JC, Liu Y, Qi X, Yap YL, Dikshit V, et al. Performance evaluation of projet multi-material jetting 3D printer. Virtual Phys Prototyping 2017;12:95–103.

[83] Pradel P, Bibb RJ, Zhu Z, Moultrie J. Exploring shape complexity factors in material deposition and material jetting build time. In: Additive manufacturing for products and applications (AMPA). Zurich, Switzerland; 2017.

[84] Baumers M, Tuck C, Dickens P, Hague R. How can material jetting systems be upgraded for more efficient multi-material additive manufacturing. In: Proceedings of the solid freeform fabrication (SFF) symposium. Texas: The University of Texas at Austin; 2014.

Materials for 3D Printing Cardiovascular Devices

3

Sanlin Robinson[1], Amir Hossein Kaboodrangi[2,3], Simon Dunham[2,3], Robert Shepherd[1]

Sibley School of Mechanical & Aerospace Engineering, Cornell University, Ithaca, NY, United States[1]; Department of Radiology, Weill Cornell Medicine, New York, NY, United States[2]; Dalio Institute of Cardiovascular Imaging, NewYork-Presbyterian Hospital, New York, NY, United States[3]

INTRODUCTION

The overarching goal of this book is to provide readers with a clear idea of the current state of the art, the challenges and limitations, and the future directions for 3D printing in cardiovascular medicine. Chapters 5–13 describe many of these applications in detail and provide a good picture of the requirements from a medical standpoint; however, many of these applications are limited by the type and quality of material that can be manufactured via 3D printing. For this reason, this chapter focuses primarily on describing the current state of the art in 3D printed materials with an emphasis on material systems and properties that are relevant to cardiovascular application. This is by no means an exhaustive review of 3D printing materials, but is designed to give readers a sense of what materials and materials properties are most relevant to cardiovascular medical applications, and what can and cannot be accomplished today with these systems, as well as future directions. It is also worth mentioning that a wide variety of highly functional models and devices can be fabricated by using replica molding based on 3D printed parts; however, this is outside of the scope of this chapter.

Because the applications as well as the materials themselves are so different, we have chosen to divide this chapter between synthetic (or acellular) and bioprinting. In the case of synthetic 3D printing, we will be considering applications such as physical models, imaging phantoms, mock flow loops, surgical guides, and training tools. In all cases, mechanical properties that match or mimic those of tissue are especially critical. For these reasons we focus primarily on 3D printing of soft materials, materials with complex mechanical response, and materials that actuate. We will also briefly consider other properties, such as cytotoxicity, hemocompatibility, and biocompatibility. In the case of bioprinted devices, the primary goal is to create devices, scaffolds, and other types of tissue or tissue-like constructs that can mimic the physiological function of the heart. Here devices like cell scaffolds, labs on a chip, organs on a chip, and cell seeded

3D Printing Applications in Cardiovascular Medicine. https://doi.org/10.1016/B978-0-12-803917-5.00003-1

implants are of particular interest. As a result, we will focus on materials to facilitate printing of cell, materials to create vasculature and other structures for perfusion and nutrient transport in tissue, and other factors in describing the types of printing that can be achieved and what considerations are most relevant for bioprinting.

SYNTHETIC MATERIALS

To date, a majority of applications in 3D printing for cardiovascular medicine have been realized with synthetic materials. Many applications such as 3D models for preprocedural visualization, surgical guides, and even basic flow models can be created readily with rigid polymers that are available today commercially in most 3D printers. We will describe these briefly; however, this information is widely available and includes case descriptions provided from a variety of 3D printing vendors. The majority of this half of the chapter will focus on synthetic polymeric materials for creating realistic models which possess the "feel and function" of real cardiac tissue. While elastic modulus (E) is an insufficient measure of these properties, it is relatively easy to characterize and provides a decent proxy for discussing which materials are of interest in creating "tissue-like" constructs. For these reasons, we will focus on soft materials, especially those with elastic modulus between 10 kPa and 100 MPa, which represent the range of mechanical properties for most cardiac tissue (with the exception of calcifications) and are almost the most challenging mechanics to obtain in a 3D printed ink or resin (many rigid materials already exist commercially). While the mechanics of cardiac tissue are highly complex (nonlinear, anisotropic, etc.) and depend on a variety of factors such as the age [1] and disease history [2] of the individual, as well as the conditions the tissue is being tested under, there are ranges of mechanical properties that can be informative in trying to create realistic cardiac models. Myocardial tissue is typically thick and soft ($E \sim 0.02{-}0.5$ MPa) and may be even stiffer for diseased tissue [3]. Arteries are slightly stiffer (~ 100 kPa-10 MPa) and vary in properties depending on their location (i.e., carotids tend to be softer than coronaries) [4]. Features like chordae tendineae and valve leaflets tend to be even stiffer, with secant modulus on the order of 100 MPa approximately, under tension [5]. Finally, calcifications, depending on their composition can range from 100 MPa to ~ 10 GPa [4]. For more detailed understanding of cardiac tissue and it's complex mechanical response, we refer readers to more thorough resources [6]. In addition, we will highlight materials, which possess anisotropic, nonlinear, or hierarchical mechanics, more similar to tissue. Finally, we will consider materials and systems, which can be mechanically actuated to create functionality similar to myocardial tissue.

COMMERCIAL MATERIAL SYSTEMS

We refer the readers to the previous chapters for a more thorough description of 3D printing technologies, however, it is worth highlighting the capabilities of the current class of commercial 3D printers. For applications where rigid polymeric materials are suitable, primarily 3D print models for visualization or preprocedural planning, many commercial materials are suitable. In the case of fused deposition modeling (FDM) printers, typical materials are acrylonitrile butadiene styrene (ABS), polylactic acid (PLA), and polyvinyl alcohol (PVA), all of which are inelastic and fairly rigid. For photocurable technologies such as polyjet (PJ) or stereolithography (SL) systems, typically resins consist of some type of acrylate/methacrylate chemistry, although resin specifics can vary. Provided a printer with sufficient resolution, all of these materials are suitable for most models and can be printed in multicolor models as well. A much smaller number of printers offer flexible materials (we define as capable of strain, $\varepsilon > 10\%$), suitable for more functional 3D print models. Thermoplastic polyurethane (PU) powders can be used with selective laser sintering (SLS) printers [7]. Urethane-acrylate systems such as Stratasys Tango materials can be printed with PJ printers, and because of their compatibility with rigid acrylate systems can be printed with a range of material properties determined by the mixing ratio of rigid and flexible components allowing for multimaterial printing. Both of the materials provide a powerful approach to creating realistic feeling models of cardiac anatomy. The multimaterial printing has been used for a wide variety of demonstrations of cardiovascular models [8−11]. Fig. 3.1 illustrates a series of models including aortic aneurysms, heart valves, and vascular models. While these represent some of the more sophisticated models achievable with commercial systems, they have several drawbacks. The materials have very limited mechanical stability when printed in thin-walled configurations below 1.5 mm; support material can be difficult to remove, and printed surfaces can have rough finishes. All of these limitations make it difficult to print realistic flow models, especially vascular models. Very recently, a new class of 3D printed Carbon Elastomeric Polyurethane and silicones printed via single-lens reflex (SLR) cameras have overcome these limitations, providing excellent durability even in thin-walled configurations (several 100 μm). Furthermore, support material removal is not required and interior walls are smooth. The primary drawback of these materials is multimaterial printing is presently not practical, so models cannot represent the full complexity of true cardiac tissue. While these systems are powerful, materials that are very soft (low strain tangent modulus, $E < 1$ MPa) possess complex mechanics, actuators, and cell-laden materials that still require more custom setups and are currently the topic of academic research. The remaining sections highlight these efforts.

ELASTOMERS

Because cardiac tissue consists of very soft materials ranging from kPa to MPa moduli, systems that provide softer mechanics and are associated with larger strains are of particular interest. Elastomers are network polymers with a low cross-link density; this molecular structure makes them highly stretchable (ultimate strains, $\varepsilon_{ult} = 100\%−1500\%$) materials that easily recover their initial shape after the

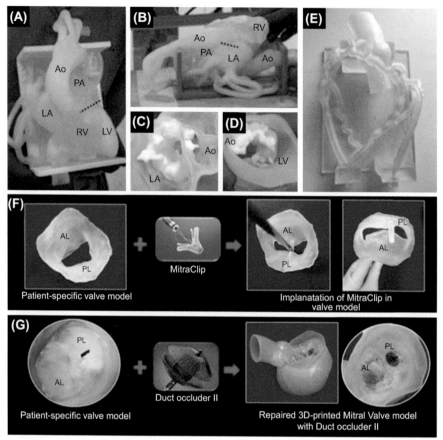

FIGURE 3.1

Patient-specific 3D print models produced from commercial 3D printers. (A—D) Full heart model from flexible materials with rigid mitral annular calcification [8]. (E) Aortic root and coronary flow model used to simulate diagnostic measurement of fractional flow reserve [10]. (F—G) 3D print valve models used to test cardiac interventions with MitraClip and Duct occlude [11].

Reproduced with permission from Izzo RL, et al. 3D printed cardiac phantom for procedural planning of a transcatheter native mitral valve replacement. Proc SPIE Int Soc Opt Eng 2016;9789; Sommer K. et al. Design optimization for accurate flow simulations in 3D printed vascular phantoms derived from computed tomography angiography. Proc SPIE Int Soc Opt Eng 2017;10138; Vukicevic M. et al. 3D printed modeling of the mitral valve for catheter-based structural interventions. Ann Biomed Eng 2017;45:508—19.

applied load is removed [12]. This class of polymers have very low glass transition temperatures ($T_g < 25°C$), making them soft and pliable at room temperature. Most elastomers used in biomedical devices are either polyurethanes, polydimethylsiloxanes (PDMSs), or hydrogels. The chemistry of the resin and the type and conditions of printing have a dramatic effect on the resultant polymer network elasticity; therefore, the choice of these materials allows the designer to span a very wide range of mechanical properties.

PUs are typically formed from reactions of polyols and isocyanates and can be formulated to yield a wide variety of properties depending on the side chains incorporated during reaction. Depending on the level of intermolecular interactions in the resultant polymer structure, these materials can be thermoplastic or thermosetting. Ninjaflex© is an FDM compatible filament comprised of thermoplastic PU (TPU) that can be melt extruded [13], although it can be challenging to produce leak-free models without producing relatively thick walls. PU particles can be printed from aqueous dispersions using direct ink writing (DIW) [14]. Methods such as stereolithography can more effectively produce closed, leakfree models, while maintaining soft mechanics (~ 2.5 MPa) and thin walls (<200 µm) [15]. In addition, provided they were produced with resins free of cytotoxic constituents, PUs can be biocompatible and integrated with cells or tissues [16]. For similar reasons, PUs are often used as components in catheter sheaths and extrusions, medical balloons, and other blood contacting medical devices.

PDMS or silicones are another common elastomeric system. Their mechanics also depend heavily on processing and polymer properties; however, they tend to be softer and more extensible than PUs. They also tend to remain elastic over a larger range of deformation, whereas PU is more prone to plastic deformation. PDMS resins are typically formed from hydrosilylation and condensation in the presence of a catalyst. For this reason, there are many two-component mixtures of liquid siloxanes that can be mixed and cured over the course of hours [17]. While melt processing is not feasible for PDMS, their resin systems are fairly compatible with DIW [18–20] and SLR [21]. For these reasons, a variety of cardiac phantoms and flow models are made from dip coating or replica molding 3D printed templates to form thin-walled, flexible models [22]. There are also a variety of commercial services available that offer these types of models, either based on idealized anatomies or customized based on patient imaging [23].

Hydrogels are a powerful material system whose mechanics can be tailored over a very wide range based on cross-link density, intermolecular bonding, and degree of swelling. As a result, resins with elastic modulus as low as several kPa [24] are achievable. One challenge is that many of these parameters are interdependent. Hydrogels are also useful vehicles for cells, which will be discussed in further detail in the Bio-inks section below. In addition, because such a wide range of hydrogels can be produced from common acrylate/methacrylate chemistries, and the ease with which these systems can be polymerized via free radical polymerization, it is very easy to create photocurable resins compatible with DIW [25] and SLR printers [26–29]. More recently, approaches based on ionic-covalent entanglements and double networks have yielded tougher hydrogels [25,28]. One drawback is the fluid permeability of materials, which may be incompatible with certain types of watertight cardiac flow phantoms. In addition, the sensitivity of these materials to hydration can make them harder to work with and may require additional encapsulation.

HIERARCHICAL MATERIALS

While the homogeneous materials described above provide mechanics similar to a variety of relevant cardiac tissue types, real tissue is much more complex, with multiscale features which yield complex, anisotropic, and hierarchical properties [6]. As an example, arterial tissue consists of three layers, intima (endothelial cell dominated), media (muscle cell dominated), and adventitia (collagen fiber dominated), each with varying mechanical properties. The difference in mechanics between layers, coupled with the directional nature of the collagen fibers in the adventitia result in asymmetric and nonlinear properties, which are important to the functioning of the arteries. Most notably, the nonlinear "J-shaped" response allows arteries to dilate under pulsatile flow, providing a capacitive response with minimal risk of tissue damage from overextension [30]. Most 3D printed systems can be utilized to create structures with complex multiscale hierarchy at macroscopic length scales through creation of complex open frame structures; however, most tissue and other biological tissues possess these features at length scales below the resolution of most 3D printers (submicron to several mm scales). As a result, we will focus primarily on unique materials that by virtue of their intrinsic properties or the way that they are processed produce multiscale, hierarchical materials.

While there are a wide variety of methods to produce 3D printed structures with complex microstructure, to date, few methods provide the control required to recapitulate complex tissue-like structures. These approaches, however, are worth considering; they demonstrate methods of control over 3D printed microstructure that with sufficient refinement may provide the control to create more tissue-like structures in the future. The simplest approach is to print composite blends with embedded micro/nano materials whose orientation is controlled during printing, either by the printing process itself or by some external stimuli. In the former case, particles and fiber can be oriented simply by the fluid forces associated with extruding an ink or resin through a narrow nozzle during jetting or extrusion [31–34]. In the latter, acoustic [35] or magnetic stimuli [36–38] are applied to ink or resins during printing to contribute structure or orientation to filler materials. These approaches can provide significant control, but require additives that are sensitive to the stimuli (i.e., magnetic nanoparticles). Alternatively, freeze casting has been used to create complex (typically lamellar) structures during printing [36]. Other approaches involve the introduction of porosity by printing phase separated polymer blends where one phase can be sacrificially removed following printing [39,40]. Utilizing this approach with elastomeric foams can create very soft mechanics for tissue-like cardiac models [41]. Finally, more direct control can be created through various forms of coextrusion of printed inks to create coaxial filaments [42]. All of these approaches provide a powerful set of methods to create more complex microstructure in 3D printed materials. As these approaches are explored with the soft materials discussed above, more tissue-like materials should be realized.

3D PRINTING ACTUATORS

One of the greatest challenges to creating highly functional cardiac models is mimicking the active actuation of cardiac tissue. The best example of this is in left ventricle, which provides sufficient contractions to reduce left ventricular blood volume to $\sim 50\%-60\%$ of its diastolic state, at every beat. Passive flow phantoms, even those with soft materials that utilize pulsatile pumps to produce physiologically realistic flows, produce unrealistic results. Because the pressure is driving the model to expand, it provides its maximum volume at systolic pressures, opposite of normal cardiac cycles. In order to develop a truly realistic phantom, materials and systems that can actively actuate are critical.

One of the simplest approaches to overcome this challenge is the use of soft robotics. Here, materials are fabricated with air- or watertight, pneumatically actuated features that, when pressurized, produce active actuation [43–45]. A powerful example of this is a cardiac flow model where sealed chambers were fabricated outside the left and right ventricles, such that when actuated, these produce contraction in the ventricular volumes [46]. Another approach has been to incorporate porous foam actuators into the ventricular wall [41]. Once again, actuation produces contractions in the ventricular volume. While these types of models produce more realistic actuation, true ventricular contractions are complex with linear and rotational components. Maps of the complex helical myocardial bands that drive these contractions reveal its underlying complexity [47]. It is likely that the most realistic models will result from embedded muscle-like actuators with geometries that match myocardial bands embedded in soft "tissue-like" ventricular walls. Recently, the ability to 3D print more complex soft robotic actuators has seen advances, with a variety of new resins and functional demonstrations being produced based on FDM, DIW, and SLR printing using polyurethanes [45,48] and hydrogels [29]. The primary drawback with soft robotic approaches is the need for pneumatic drivelines and controls, which can limit portability and add complexity to the system (Fig. 3.2).

There are a variety of materials that produce actuation through an intrinsic, mechanical response to external stimuli. The simplest examples are hydrogels and other materials that respond to swelling. A variety of 3D printed models have been shown to produce complex mechanical response to swelling based on this phenomenon [49–51]. Similarly, 3D printed thermoresponsive materials based on acrylate-based shape memory materials [52], liquid crystal elastomers [53], and volatile liquid embedded silicones [54] have been demonstrated. All of these systems can show a dramatic actuation response. The drawback of both of these approaches is that they lack the temporal response to mimic cardiac contractions. A notable exception may be in thin-walled models with continuous fluid flow through the model. Here the limited thermal mass and rapid convective cooling could produce sufficiently rapid actuation. Electrically driven actuators have the potential to overcome these limitations, however, these provide their own limitations. Most provide limited

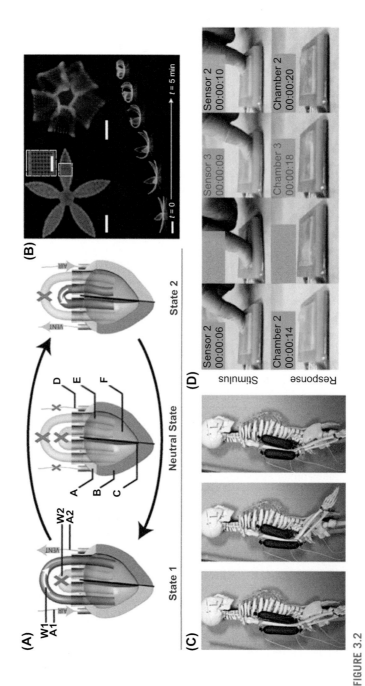

FIGURE 3.2

3D printed actuators. (A) Soft robotic cardiac model based on pneumatically expanding ventricular walls [41], (B) 4D biomimetic materials based on hydrogel composite ink. Swelling induces complex changes in shape programmed into the material based on the orientation of cellulose fibrils [51], (C) 3D printed thermal actuators based on 3D printed ethanol embedded elastomers. Upon heating, the ethanol evaporates creating local expansion [54], (D) 3D printed dielectric elastomeric actuator array. Finger pushes are detectible based on changes in capacitance of sensors [59].

Reproduced with permission from Mac Murray BC, et al. Poroelastic foams for simple fabrication of complex soft robots. Adv Mater 2015;27:6334—40; Sydney Gladman A, et al. 4D printing. Nat Mater 2016;15:413; Miriyev A, et al. Soft material for soft actuators. Nat Commun 2017;8:596; Robinson SS, et al. Integrated soft sensors and elastomeric actuators for tactile machines with kinesthetic sense. Extreme Mech Lett 2015;5:47—53.

actuation force. Beyond this, most require either high currents or high voltages to operate, as well as 3D printed stretchable conductors to provide local electric fields. A variety of metal nanoparticle loaded elastomers [55] and conductive hydrogels [24] meet this requirement. Piezoelectric materials provide only limited displacements [56]. Dielectric elastomeric actuators can generate very large displacements and can be used to sense their own deformation, but provide limited forces, are limited to thin configurations, and require very large voltages to actuate [57–59]. 3D printed ionic polymer-metal composites provide low voltage actuation, but require complex liquids for actuation and are limited to relatively thin configurations [60]. While there are a variety of promising approaches, currently, they all possess critical limitations.

BIOPRINTING MATERIALS

Bioprinting is the precise, layer-by-layer patterning of cell suspensions, cell-laden/ acellular hydrogels [61–63], cell spheroids [64,65], or polymeric inks [66–68] for the fabrication of tissue constructs, organ modules, or organ-on-a-chip devices, with the grand challenge of printing fully functional organs [69–71]. The field of bioprinting is typically applied to two major areas of research, either the repair/regeneration of tissues and organs, or the creation of in vitro models for the study of biological development, disease progression, and drug interaction [72]. While the library of bioprinting materials has greatly expanded in the last 15 years, there are still significant technical hurdles surrounding the 3D printing of viable tissue constructs. Some of these challenges include (1) patterning multiple cell types in sufficient resolution and density, (2) re-creating an extracellular matrix (ECM) that delivers physiologically relevant mechanical and chemical signals to cells, and (3) incorporating vasculature to deliver nutrients and oxygen, remove metabolic waste, and allow the printed tissue construct to anastomose in vivo. In the following sections we will describe the most common materials and strategies employed in bioprinting and the material properties that should be considered when selecting a bioink.

BIOINK PROPERTIES

When selecting a bioink for fabricating tissue constructs, several factors must be considered in order to satisfy material, manufacturing, and biological constraints. Specifically, the ideal bioink should be optimized for printability, biocompatibility, degradation kinetics, mechanical properties, swelling, and biomimicry [70,72,73]. Other properties—depending on the final application—should also be considered, such as hemocompatibility and degradation by-products.

Printability of a bioink is determined by its processability and print fidelity—or the ability of the ink to maintain its structure after being printed. Since there is a large variety of printing technologies available, the bioink's viscosity, surface tension, and cross-linking mechanisms have to be suitable for the chosen printing method. The main techniques used in bioprinting are inkjet printing, direct-ink-write (also known as extrusion-based printing), and laser-assisted printing. Inkjet printing is the ejection of multiple, tiny (1—100 pL) droplets onto a surface that then coalesce to form a printed layer. Bioinks used in inkjet printing must possess very low viscosities ($\eta < 10$ mPa s), be able to withstand high shear rates $\left(\dot{\gamma} \sim 10^5 - 10^6 \text{ s}^{-1}\right)$, and gel immediately after exiting the nozzle [72,74,75]. These inks should not gel before leaving the print head as this will cause blockages; however, if they do not cure within a reasonable amount of time then the printed structure will lose its intended shape [75]. Materials that have been printed using this technique are hydrogels, colloidal suspensions, and cell suspensions [72]. DIW is an extrusion printing method, where highly viscous ($\eta \sim 30 - 6 \times 10^7$ mPa s) inks are pneumatically or mechanically pushed through a nozzle (diameter $\sim 50-300$ μm) [69,72,75]. It is important to optimize the material's viscosity for DIW. Low viscosity results in low shape fidelity of the printed structure after deposition; however, high viscosity results in a restrictive environment for cell migration and proliferation post printing [75]. This method of printing is highly versatile as many materials can be formulated to fit these requirements. When printing cell-laden inks, very high cell densities are attainable; however, the print resolution is poor (200—1000 μm) and cell viability is low due to the high shear stress experienced at the nozzle [75]. One way to overcome this shear stress limitation is by increasing the size of dispensing nozzle at the cost of resolution [76]. Another option is to take advantage of non-Newtonian fluids with shear-thinning behavior. For example, sodium alginate possesses a strong shear-thinning behavior; its viscosity decreases with increasing shear rate in the nozzle during printing, but sharply increases upon deposition allowing it to maintain its printed structure [76]. DIW has been used to print polymer solutions, hydrogels, colloidal suspensions, decellularized extracellular matrices, and cell spheroids [69,72]. Laser-assisted printing uses bioinks of moderate viscosities (1—300 mPa s), and since this method is nozzle-free, it avoids the issues of clogging and high shear stress that would otherwise occur at the nozzle if using DIW [75]. Laser-assisted printing works best with biomaterials that are able to cross-link very rapidly. It has been demonstrated to work well with ionically cross-linked alginates and enzymatically cross-linked fibrinogen [75].

The biomaterials we will describe are either physically or chemically cross-linked. Physical cross-linking is nonpermanent; the interactions involved in the gelation of the polymer can be broken and reformed several times. The bonds formed in these crosslinks can be made via ionic interactions, van der Waal's forces, hydrogen bonding, or high molecular weight chain entanglements [72,75]. Physical cross-linking is useful for creating a structured network for DIW printing. The network structure is often disrupted at high shear stresses allowing the polymer to easily flow through a microsized nozzle; the structure then reforms at low shear permitting the ink to maintain its printed structure for a short period of time. For printed structures to be mechanically robust over longer periods of time, the use of materials with permanent, chemical cross-linking is recommended. Chemical cross-linking happens when covalent bonds are formed between the polymer strands/networks; these bonds can be formed through photopolymerization, thermal polymerization, Michael-type addition reactions, click chemistry, enzymatic reactions, or a change in pH [72,75].

Native soft tissues and organs have been reported with Young's moduli ranging from 0.1 kPa to 1 Mpa [77]. When engineering tissues, it is critical to recapitulate the native mechanics of the microenvironment as cells draw important regulatory cues (i.e., adhesion, migration, proliferation, differentiation) from the stiffness of their surroundings [78]. For example, when cardiomyocytes are cultured on synthetic hydrogels with physiological stiffness (polyacrylamide, $E \sim 10$ kPa), they display striated myofibrils and good contractility; however, when they are cultured on very stiff substrates they lose their in vivo phenotype and no longer contract rhythmically [78]. Additionally, mechanical properties of hydrogels have demonstrated a critical role in growth and differentiation of stem cells. For example, human mesenchymal stem cells differentiate into neurogenic cells when they are cultured in hydrogels of $E \sim 0.1-1$ kPa stiffness, into myogenic and osteogenic cells when placed in $E \sim 8-17$ kPa, and into nascent bone when placed in $E > 34$ kPa [77]. We can mimic a variety of soft tissues by tuning a polymer's tangent modulus profile through manipulation of the chemistry, polymer concentrations, and degree of cross-linking [76,77,79]. For example, researchers have shown that the Young's moduli of polyethylene glycol (PEG)-based hydrogels can be varied from $E \sim 60-500$ kPa by increasing the polymer concentration from $\phi \sim 10\%-20\%$ [80]. Another study demonstrated that by increasing the degree of cross-linking in hyaluronic acid-tyramine hydrogel, a concomitant increase in its stiffness ($E = 5.4-11.8$ kPa) was observed [78].

BIOINKS

The available library of bioprinting materials is still evolving, it can range from cell-laden hydrogels [61,63,66], hydrogel-free cell suspensions, cell/tissue spheroids [64,65,81], decellularized ECM [82,83], to acellular polymers [62,67,68,71,79]. Acellular inks can be processed using cytotoxic reagents, high temperatures, or methods involving high shear stress, as they do not need to be cell-compatible until well after printing and post processing [84]. In contrast, cell-containing bioinks have more stringent biocompatibility and processing requirements, because they must always provide a suitable environment for living cells. Hydrogels are typically used in bioprinting due to their ease of modification for different printing modalities and wide variety of available materials [72]. Both synthetic and naturally derived hydrogels have been used as cell-laden inks; these macromolecules have a hydrophilic polymer backbone that swells in water. Hydrogels are an attractive material for cell-encapsulation applications because of their high water content and mechanical properties; additionally, we can use them to fabricate three-dimensional microenvironments that recapitulate native ECM [76]. The following sections will discuss the most common materials used to formulate bioinks. We have chosen to split this discussion into two categories, naturally derived hydrogels and synthetic polymers.

Natural Bioinks

Naturally derived polymers are typically polysaccharide chains, which may also contain peptides and proteins. These polymers are particularly useful in bioengineering because of their inherent bioactivity [72]. The most common natural hydrogels are alginate, gelatin [85,86], collagen [87,88], fibrin [72,89], and hyaluronic acid [72,89]. These hydrogels are superior to their synthetic surrogates for tissue engineering applications, because they provide a suitable ECM and biological factors for cells to grow, proliferate, and migrate.

Alginate is extracted from algae or seaweed, making it very inexpensive, abundant, biocompatible, biodegradable, and nontoxic [72,79,90]. Upon exposure to calcium chloride, alginate rapidly forms a physical gel through ionic interactions with calcium (Ca^{2+}) [62]. By flushing the alginate hydrogel with a chelating agent, such as sodium citrate or ethylenediaminetetraacetic acid, the Ca^{2+} can be removed and the gel will liquefy; this could be useful for creating open vasculature and sacrificial templates. Alginate—while biocompatible—is not bioactive; therefore, it must be blended with other natural polymers in order for cells to adhere and proliferate [79]. Duan et al. formulated alginate/gelatin blends and encapsulated porcine aortic valve interstitial cells and smooth muscle cells in order to bioprint heterogeneous aortic valve constructs [91]. They mixed gelatin and alginate in a sodium chloride solution at concentrations of 0.06 and 0.05 g/mL respectively. They used a dual-print head system to print the alginate/gelatin blend into a valve conduit; the structure was able to maintain its shape until being cross-linked using calcium chloride. The encapsulated cells were >80% viable after 7 days of culture [91]. Khalil and Sun prepared sodium alginate solutions of varying concentrations (1%–3% w/v); they mixed the solutions with rat heart endothelial cells at 500,000 cells/mL cell density [92]. They extruded this mixture through a 250 μm nozzle into a 0.5% (w/v) calcium chloride solution where it gelled, thereby encapsulating the cells. They found that the endothelial cells proliferate well within this range of concentration [92].

Collagen is one of the main components of natural ECM; its isolation and purification protocols are well established, and therefore it is a common material among industry and research institutions [79]. Despite the fact that it is naturally derived, pure collagen is not an ideal bioink, because it may trigger undesirable biological signaling when used without additional ECM components. Therefore hybrid materials that contain collagen, as well as other ECM materials like elastin, glycosaminoglycans, fibrinogen, and laminin, may be more ideal for tissue engineering [79]. In order to fabricate muscular patches, Moon et al. coated smooth muscle cells in collagen and then printed them in droplets via inkjet printing

[93]. Using a collagen bioink that is liquid at low pH (pH ∼ 4.5) and gels at neutral pH, and a thermally reversible gelatin hydrogel, Lee et al. generated collagen scaffolding with embedded channels [88]. They used gelatin as a sacrificial ink to create open channels, and sodium bicarbonate in water as a cross-linker for the collagen hydrogel precursor.

Gelatin is produced through the partial hydrolysis of collagen's triple helix structure into smaller molecules. Below 30−35°C, gelatin is a physically cross-linked solid; since it melts into a flowable liquid at physiological temperatures, permanent, chemical cross-linking is required for structures composed from gelatin [79,90]. Physically gelled gelatin has been used as a supportive material during printing heart valve conduits [94] and vasculature [88,95]. Gelatin has been enzymatically cross-linked with fibrin using transglutaminase to recapitulate tissue ECM [66,67]. Methacrylated gelatin (GelMA) is commonly used among researchers to formulate bioinks that are bioactive, mechanically tunable, and photocross-linkable [61,96−99].

Fibrin is a fibrous protein that is formed by the enzymatic cleavage of fibrinogen by the protease thrombin; in vivo, this cross-linking mechanism is critical during clot formation and wound healing [61,66,79,85,96,97,100−103]. Tissue engineers have taken advantage of this rapid cross-linking to create hydrogels that are mechanically robust, and representative of natural ECM [66,67,104,105].

Hyaluronic acid (HA) is another major component of native ECM; it is a non-sulfated glycosaminoglycan consisting of repeating disaccharide units. Since these units are highly polar, they attract water, making them great lubricants in the body [79,90]. By modifying the HA backbone with methacrylate groups, it is possible to form a photocross-linkable hydrogel [79,101,106]. Duan and colleagues investigated how tuning the stiffness of methacrylated HA/GelMA blends affects valve interstitial cells' proliferation, spreading, and phenotype [106]. Alternatively, HA can be modified by adding thiol functional groups [107]; this allows Michael-type addition cross-linking with polyethylene glycol-diacrylate [108]. Gold nanoparticles have also been used to dynamically cross-link thiol-modified HA; the extruded HA bioink was initially cross-linked into a filament, but over time the printed layers fused together [108]. Recently, the Burdick lab has been exploring embedded printing using HA-based hydrogels that cross-link via cyclodextrin-adamantane host-guest interactions [107,109,110]. In their work, they print a shear-thinning HA hydrogel directly into a supportive hydrogel bed. After printing, the supportive hydrogel is able to self-heal and self-assemble by physically bonding with the printed gel.

Synthetic Bioinks

Many synthetic materials were discussed in previous sections of this chapter, but most of the mentioned materials were not intended to house cells or tissues; therefore, we have created this subsection in order to discuss synthetic polymers that are used specifically for tissue engineering. While some synthetic hydrogels can encapsulate cells while being bioprinted, many of these synthetic materials are processed at high temperatures, or use toxic solvents during printing, and therefore cells are typically seeded on the structures post printing. Although synthetic materials do not have the same biocompatibility as the natural hydrogels discussed above, the ease of control over their molecular weight, molecular weight distribution, and cross-linking densities allows mechanical modifications (i.e., elastic modulus) to suit a particular application, making them a very attractive option for bioprinting materials [70,79]. Synthetic materials are biologically inert, meaning, they lack specific binding sites for cell adhesion; therefore, they should be combined with functionalized or naturally derived materials that carry specific biological factors that simulate the ECM [72].

Polycaprolactone (PCL) is a popular synthetic polymer in engineering scaffolding for tissues, because it is biocompatible and provides robust mechanical support. This polyester-based polymer melts at a relatively low temperature of 60°C, and is biodegradable. Since PCL lacks specific binding sites for cell adhesion, it is often combined with bioactive materials to promote adhesion and proliferation of cells [79]. Since PCL cools quickly, the Atala group has explored printing it in conjunction with natural hydrogels (gelatin, HA, fibrin) to create cell-laden tissue constructs. They demonstrated a multinozzle extrusion printer that can simultaneously print cell-laden hydrogels, supporting PCL, and sacrificial pluronic hydrogel [111]. They chose to use PCL because it takes ~ 2 years to degrade, and therefore can provide long-term mechanical support to the printed tissues/organ constructs (Fig. 3.3).

Common synthetic hydrogels include PEG-based materials, polyacrylamide, and pluronic. The Lewis lab has demonstrated using pluronic as a fugitive ink; its gel-to-liquid transition below 10°C is particularly useful for creating scaffolds that are later removed to create open, perfusable networks; we discuss this polymer in detail in the bioprinting vasculature section below [66,67]. In order to make it photopolymerizable, PEG is typically chemically modified into PEG-diacrylate (PEG-DA) or PEG-methacrylate [79]. For example, in order to replicate the mechanical heterogeneity found in native heart valves, Hockaday et al. varied the concentrations and molecular weights of PEG-DA; they blended in alginate to modify the viscosity for printing [94]. To simulate the stiffness of the aortic root, they used 20% of 700 MW PEG-DA blended with 12.5% w/v alginate. Since valve leaflets are very extensible and compliant, they blended 5% 700 MW PEG-DA and 7.5% 8000 MW PEG-DA with 15% w/v alginate. Since the root and leaflets were formulated using the same base materials and cross-linker, good adhesion was seen between printed layers [94].

BIOPRINTING VASCULATURE

Without adequate perfusion of oxygen and nutrients, engineered tissues will quickly become necrotic; living cells must be within a few hundred microns of a perfused channel in order to remain viable [66,68]. When engineering tissues, researchers must consider the diffusion requirements of the construct as a whole and the unique

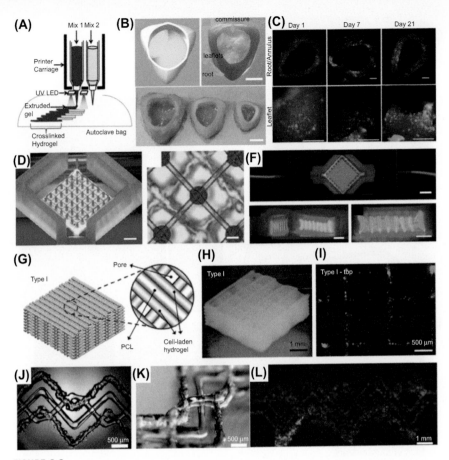

FIGURE 3.3

(A) Schematic of multinozzle 3D printer used to print two different hydrogels. Printing within sterile autoclave bag helped maintain aseptic conditions. (B) Asymmetric valve model printed with two blends of hydrogels. Aortic root printed using stiff PEG-DA formulation 20% 700 MW PEG-DA with 12.5% alginate, the valve leaflets were printed using a more compliant hydrogel of 5% 700 MW and 7.5% 8000 MW with 15% alginate, ID—22, 17, and 12 mm (scale bars = 1 cm). (C) Live/Dead imaging of porcine aortic valve interstitial cells on the root and leaflet of printed valve constructs [94]. (D) Image of 3D printed, thick tissue construct with surrounding silicone gasket. Red corresponds to the pluronic fugitive ink used to create vasculature (scale bar = 2 mm) (E) Bright-field image of printed construct, pink lines are printed pluronic ink. Irregular lines are cell-laden ink (scale bar = 50 μm). (F) Top down image of tissue construct on chip—red is demonstrating flow of media, while green is displaying cell-laden inks. Cross-sectional views of thick tissue construct (scale bars = 5 mm) [66]. (G) Schematic of tissue construct, PCL (*gray*), cell-laden hydrogel (*green and red*), and empty spaces. (H and I) Photo and fluorescent image of printed tissue construct [111]. (J) Bright-field image of three bioinks (two cell-laden inks and pluronic ink in center). (K) Side view of printed cell-laden and pluronic inks. (L) Fluorescent image of fibroblasts (*blue*), human neonatal dermal fibroblasts (*green*), human umbilical vein endothelial cells (*red*) [63].

Reproduced with permission from Hockaday LA, et al. Rapid 3D printing of anatomically accurate and mechanically heterogeneous aortic valve hydrogel scaffolds. Biofabrication 2012;4:035005; Kolesky DB, et al. Three-dimensional bioprinting of thick vascularized tissues. Proc Natl Acad Sci USA. 2016;113:3179–84; Kang HW, et al. A 3D bioprinting system to produce human-scale tissue constructs with structural integrity. Nat Biotechnol 2016;34:312–9; Kolesky DB, et al. 3D bioprinting of vascularized, heterogeneous cell-laden tissue constructs. Adv Mater 2014;26:3124–30.

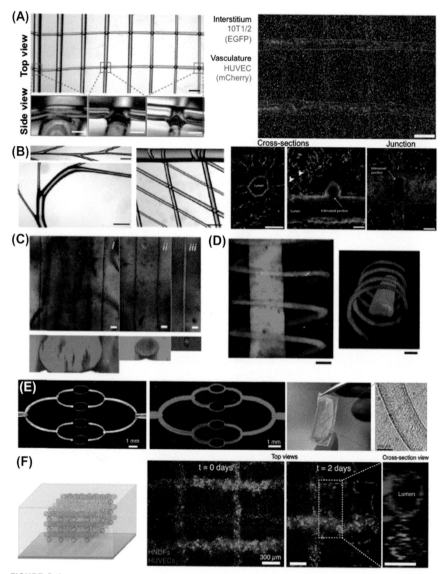

FIGURE 3.4

(A) Left, printed carbohydrate glass lattice of varying size filaments imaged from top and side (scale bars = 200 μm). Right, after encasing the printed glass in fibrin gel, the glass was removed and then human umbilical vein endothelial cells (HUVECs, expressing mCherry) were seeded throughout vascular network and human fibroblasts (10T1/2, expressing EGFP) throughout interstitium (scale bar = 1 mm). (B) Left, printed carbohydrate glass in various architectures, multiscaled curved y - junctions and angled filaments (scale bar = 1 mm). Cross-section of open lumen after 9 days in culture, needs of the targeted cell/tissue type [112]. Several methods to create networks of vasculature have been explored, such as (1) self-assembly of patterned cells, (2) controlled release of growth factors to promote vasculogenesis, (3) printing fugitive inks, and (4) coaxial nozzle printing. Researchers have shown that patterning endothelial cells/spheroids [113] or proangiogenic factors [114] in engineered tissues will encourage the growth of vessels. Norotte et al. fabricated small-diameter vessels by printing multicellular spheroids and agarose rods; the spheroids fused within 5–7 days [115]. These methods take several days to generate vessels with perfusable lumens; therefore, they may not be viable solutions to meet the immediate metabolic needs of the cells in the tissue construct [112].

More promising methods, for producing large tissue constructs that remain viable over long periods of time, are those that allow for the immediate perfusion of channels [66,112]. 3D printing of fugitive inks has been explored in bioprinting to create a sacrificial template that is later removed to form a perfusable network of open lumens—mimicking in vivo vasculature [66–68]. Typically the fugitive ink is first printed into the desired architecture, then a secondary material is cast over the sacrificial network, and finally the fugitive ink is removed leaving behind a network of open lumens within the bulk material. These open lumens are then directly perfused with endothelial cells that adhere to the walls of the lumen to create a confluent monolayer, like that seen in vivo. For fugitive inks to be compatible with bioprinting, the ink (1) should support itself and span gaps when printed in, on its own (i.e., in open air), (2) must rapidly dissolve and easily flow out of the cast matrix, and (3) should not leave behind any residue, or this residue must be biocompatible with cells. Fugitive inks have been made from several materials including carbohydrate glass [68,116], pluronic [66,67], and cell-laden gelatin [95] (Fig. 3.4).

A biocompatible carbohydrate glass ink was created by mixing glucose, sucrose, and dextran in water [68]; an alternate mixture was formulated with isomalt, dextran, and water [116]. This solution was heated to 165°C to remove majority of the water, leaving a very viscous liquid glass that could be 3D printed via thermal extrusion and

---

intersecting channels form junctions that are endothelialized with HUVECs (scale bar = 200 μm) [68]. (C) Confocal images of multiscale rods printed within supportive gel through 20, 27, and 34 gauge nozzles via GHost writing of modified hyaluronic acid (scale bars = 100 μm). (D) Confocal images of printed rod with surrounding spiral printed within supportive gel using the same method (scale bars = 200 μm) [109]. (E) Printed pluronic ink within ECM matrix, pluronic is evacuated leaving behind empty channels, which are then perfused with media (*red*). Photograph of printed microvessel within bulk matrix, vascular network perfused with HUVECs (F) Schematic of tissue scaffold. Red filaments are HUVECs and green filaments are human neonatal dermal fibroblasts GelMA inks [63].

Reproduced with permission from Miller JS, et al. Rapid casting of patterned vascular networks for perfusable engineered three-dimensional tissues. Nat Mater 2012;11:768–74; Highley CB, et al. Direct 3D printing of shear-thinning hydrogels into self-healing hydrogels. Adv Mater 2015;27:5075–9; Kolesky DB, et al. 3D bioprinting of vascularized, heterogeneous cell-laden tissue constructs. Adv Mater 2014;26:3124–30.

fiber drawing [68]. The printed glass is optically transparent making it compatible with matrices that require photopolymerization for gelation. One of the major advantages of this channel forming method is that it is compatible with a wide variety of ECM materials, such as agarose, alginate, PEG hydrogels, thrombin-polymerized fibrin, and Matrigel [68]. Before infilling the glass filaments with an ECM material, the lattice is first coated with a layer of poly(D-lactide-co-glycolide) to ease the removal of the dissolved carbohydrates.

Pluronic is an aqueous block copolymer that has a thermally reversible gelation above its critical micelle concentration (CMC ∼21%) that can be exploited for creating bioinks or fugitive inks. Pluronic F127 is typically used in bioprinting applications; it has a hydrophobic poly(propyleneoxide) (PPO) block sandwiched between two hydrophilic poly(ethylene oxide) (PEO) blocks (PEO-PPO-PEO) [63]. Below pluronic's gelation temperature, this polymer solution is liquefied and flows easily. Above this temperature the PPO segments fold into the micelle core, while the PEO segments assemble into the outer shell of the micelle. The gelation temperature can be adjusted from 10 to 4°C by increasing the concentration of the polymer in water [63]. These inks are typically printed into the desired vascular structure at room temperature, after the structure has been infilled with an ECM material; the temperature is then dropped and gentle vacuum is applied to remove the liquefied pluronic [63]. The open lumens can then be perfused with endothelial cells, or other cell types depending on the application [63,66,67].

A method for including endothelial cells directly in the printed vasculature is to use cell-laden gelatin as a fugitive ink. Lee et al. have demonstrated that a 1:1 mixture of 20% gelatin and human umbilical vein endothelial cells (HUVECs) can be printed within a collagen scaffold [68,95]. After the cell-laden gelatin was fully sealed within collagen, it was heated to 37°C and liquefied. The endothelial cells were attached to the inner walls of the channel by flipping over the construct every 10−15 min. The construct was finally connected to a pump and media was made to flow gently through the channel to push out the gelatin and create perfusable vasculature lined with HUVECs.

Coaxial nozzle printing is when the tip of an extrusion printer is modified to push multiple, circumferentially layered materials out of the same nozzle [117,118]. Hollow alginate channels were fabricated by dispensing a 2% sodium alginate solution through the outer tube of the nozzle and simultaneously dispensing a 4% calcium chloride solution down the center tube of the nozzle [117]. When the calcium came into contact with the alginate solution, it diffused through the ink and caused it to physically cross-link. The outer walls of the filament remained ungelled; this allowed subsequent printed layers to coalesce with previous layers. The entire construct was slowly lowered into a bath of calcium chloride solution in order to cross-link the outer walls of the channels [117]. This process resulted in hollow filaments with inner diameters of ∼892, and ∼200 μm wall thickness. Another group took advantage of coaxial nozzle printing by printing both synthetic and natural biomaterials together. Jia et al. used a trilayered nozzle to print a blend of GelMA, sodium alginate, and 4-arm poly(ethylene glycol)-tetra-acrylate (PEGTA) sandwiched between two layers of calcium chloride solution [117,118]. This method was based on ionic cross-linking of the alginate from the calcium delivered through the core and sheath layers. The GelMA and PEGTA are then photopolymerized to form a permanent covalently bonded structure [118]. The biologically inert alginate is then removed by soaking the construct in a Ca^{2+} chelating solution; this resulted in improved spreading and proliferation of cells in the printed tissue [118].

CONCLUSION

Currently, the capabilities of commercial 3D printers provide a powerful basis for creating basic cardiac models which allow for visualization; making of surgical guides, preprocedural planning tools, and basic phantom flow loops. Current materials research, however, provides avenues for much more sophisticated models which can replicate tissue mechanics, provide actuation, and incorporate complex tissue constructs. These advances should dramatically expand the scope of cardiac 3D printing allowing for complex, highly realistic flow models for device testing and clinical training without requiring the use of animal testing or cadavers. Furthermore, bioprinting provides the promise to allow for realistic tissue models for drug testing and other complex biological response and eventually 3D printed organs. However, in order to realize these opportunities, numerous advances in material engineering and dramatically more controlled and refined printers are required.

REFERENCES

[1] Ozolanta I, Tetere G, Purinya B, Kasyanov V. Changes in the mechanical properties, biochemical contents and wall structure of the human coronary arteries with age and sex. Med Eng Phys 1998;20(7):523—33. Epub 1998/12/01. PMID:9832028.

[2] Nagueh SF, Shah G, Wu Y, Torre-Amione G, King NM, Lahmers S, Witt CC, Becker K, Labeit S, Granzier HL. Altered titin expression, myocardial stiffness, and left ventricular function in patients with dilated cardiomyopathy. Circulation 2004; 110(2):155—62. https://doi.org/10.1161/01.cir.0000135591.37759.af. Epub 2004/07/09, 15238456.

[3] Chen Q-Z, Bismarck A, Hansen U, Junaid S, Tran MQ, Harding SE, Ali NN, Boccaccini AR. Characterisation of a soft elastomer poly(glycerol sebacate) designed to match the mechanical properties of myocardial tissue. Biomaterials 2008;29(1): 47—57. https://doi.org/10.1016/j.biomaterials.2007.09.010.

[4] Akyildiz AC, Speelman L, Gijsen FJ. Mechanical properties of human atherosclerotic intima tissue. J Biomech 2014;47(4):773—83. https://doi.org/10.1016/j.jbiomech.2014.01.019. Epub 2014/02/18, 24529360.

[5] Zuo K, Pham T, Li K, Martin C, He Z, Sun W. Characterization of biomechanical properties of aged human and ovine mitral valve chordae tendineae. J Mech Behav Biomed Mater 2016;62:607—18. https://doi.org/10.1016/j.jmbbm.2016.05.034. Epub 2016/06/18. PMCID:PMCPmc5058331, 27315372.

[6] Biomechanics McEJ. Mechanical properties of living tissues by Y. C. Fung. Med Phys 1982;9(5):788−9. https://doi.org/10.1118/1.595186.

[7] Materialize. Laser sintering materials properties. 2013.

[8] Izzo RL, O'Hara RP, Iyer V, Hansen R, Meess KM, Nagesh SVS, Rudin S, Siddiqui AH, Springer M, Ionita CN. 3D printed cardiac phantom for procedural planning of a transcatheter native mitral valve replacement. Proc SPIE Int Soc Opt Eng 2016:9789. https://doi.org/10.1117/12.2216952. Epub 2016/02/27. PMCID: PMCPmc5467736, 28615797.

[9] Meess KM, Izzo RL, Dryjski ML, Curl RE, Harris LM, Springer M, Siddiqui AH, Rudin S, Ionita CN. 3D printed abdominal aortic aneurysm phantom for image guided surgical planning with a patient specific fenestrated endovascular graft system. Proc SPIE Int Soc Opt Eng 2017:10138. https://doi.org/10.1117/12.2253902. Epub 2017/06/24. PMCID:PMCPmc5476205, 28638171.

[10] Sommer K, Izzo RL, Shepard L, Podgorsak AR, Rudin S, Siddiqui AH, Wilson MF, Angel E, Said Z, Springer M, Ionita CN. Design optimization for accurate flow simulations in 3D printed vascular phantoms derived from computed tomography angiography. Proc SPIE Int Soc Opt Eng 2017:10138. https://doi.org/10.1117/12.2253711. Epub 2017/07/01. PMCID:PMCPmc5485824, 28663663.

[11] Vukicevic M, Puperi DS, Jane Grande-Allen K, Little SH. 3D printed modeling of the mitral valve for catheter-based structural interventions. Ann Biomed Eng 2017;45(2):508−19. https://doi.org/10.1007/s10439-016-1676-5. Epub 2016/06/22, 27324801.

[12] Young RJ, Lovell PA. Introduction to polymers. 3rd ed. Taylor & Francis; 2011.

[13] Ninjatek, NinjaFlex ® 3D Printing Filament. Flexible polyurethane material for FDM printers. 2016.

[14] Hung KC, Tseng CS, Hsu SH. Synthesis and 3D printing of biodegradable polyurethane elastomer by a water-based process for cartilage tissue engineering applications. Adv Healthc Mater 2014;3(10):1578−87. https://doi.org/10.1002/adhm.201400018. PMID:WOS:000343798800007.

[15] Sinh LH, Harri K, Marjo L, Minna M, Luong ND, Jurgen W, Torsten W, Matthias S, Jukka S. Novel photo-curable polyurethane resin for stereolithography. RSC Adv 2016;6(56):50706−9. https://doi.org/10.1039/C6RA05045J.

[16] Janik H, Marzec M. A review: fabrication of porous polyurethane scaffolds. Mater Sci Eng C Mater Biol Appl. 2015;48:586−91. https://doi.org/10.1016/j.msec.2014.12.037. PMID:WOS:000348749200073.

[17] Mark JE. Overview of siloxane polymers. In: Silicones and silicone-modified materials. American Chemical Society; 2000. p. 1−10.

[18] Qin Z, Compton BG, Lewis JA, Buehler MJ. Structural optimization of 3D-printed synthetic spider webs for high strength. Nat Commun 2015;6:7. https://doi.org/10.1038/ncomms8038. PMID:WOS:000355530400002.

[19] Hinton TJ, Hudson A, Pusch K, Lee A, Feinberg AW. 3D printing PDMS elastomer in a hydrophilic support bath via freeform reversible embedding. ACS Biomater Sci Eng 2016;2(10):1781−6. https://doi.org/10.1021/acsbiomaterials.6b00170.

[20] Plott J, Shih A. The extrusion-based additive manufacturing of moisture-cured silicone elastomer with minimal void for pneumatic actuators. Addit Manuf 2017;17:1−14. https://doi.org/10.1016/j.addma.2017.06.009.

[21] Patel DK, Sakhaei AH, Layani M, Zhang B, Ge Q, Magdassi S. Highly stretchable and UV curable elastomers for digital light processing based 3D printing. Adv Mater 2017;29(15):1606000. https://doi.org/10.1002/adma.201606000.

[22] Swailes NE, MacDonald ME, Frayne R. Dynamic phantom with heart, lung, and blood motion for initial validation of MRI techniques. J Magn Reson Imag 2011;34(4): 941−6. https://doi.org/10.1002/jmri.22688.

[23] Shelley Medical Imaging Technologies. Available from: http://www.simutec.com/.

[24] Odent J, Wallin TJ, Pan W, Kruemplestaedter K, Shepherd RF, Giannelis EP. Highly elastic, transparent, and conductive 3D-printed ionic composite hydrogels. Adv Funct Mater 2017;27(33):1701807. https://doi.org/10.1002/adfm.201701807.

[25] Bakarich SE, Mih P, Beirne S, Wallace GG, Spinks GM. Extrusion printing of ionic-covalent entanglement hydrogels with high toughness. J Mater Chem B 2013;1(38): 4939−46. https://doi.org/10.1039/C3TB21159B.

[26] Bakarich SE, Gorkin R, in het Panhuis M, Spinks GM. Three-dimensional printing fiber reinforced hydrogel composites. ACS Appl Mater Interfaces 2014;6(18): 15998−6006. https://doi.org/10.1021/am503878d.

[27] Bakarich SE, Gorkin Iii R, Gately R, Naficy S, in het Panhuis M, Spinks GM. 3D printing of tough hydrogel composites with spatially varying materials properties. Addit Manuf 2017;14:24−30. https://doi.org/10.1016/j.addma.2016.12.003.

[28] Valentin TM, Leggett SE, Chen P-Y, Sodhi JK, Stephens LH, McClintock HD, Sim JY, Wong IY. Stereolithographic printing of ionically-crosslinked alginate hydrogels for degradable biomaterials and microfluidics. Lab Chip 2017;17(20):3474−88. https://doi.org/10.1039/C7LC00694B.

[29] Wallin TJ, Pikul JH, Bodkhe S, Peele BN, Mac Murray BC, Therriault D, McEnerney BW, Dillon RP, Giannelis EP, Shepherd RF. Click chemistry stereolithography for soft robots that self-heal. J Mater Chem B 2017;5(31):6249−55. https://doi.org/10.1039/C7TB01605K.

[30] Shadwick RE. Mechanical design in arteries. J Exp Biol 1999;202(Pt 23):3305−13. Epub 1999/11/24, 10562513.

[31] Compton BG, Lewis JA. 3D-Printing of lightweight cellular composites. Adv Mater 2014; 26(34):5930. https://doi.org/10.1002/adma.201401804. PMID:WOS:000342147400003.

[32] Carrico JD, Leang KK, editors. Fused filament 3D printing of ionic polymer-metal composites for soft robotics. SPIE smart structures and materials + nondestructive evaluation and health monitoring. SPIE; 2017.

[33] Parandoush P, Lin D. A review on additive manufacturing of polymer-fiber composites. Compos Struct 2017;182:36−53. https://doi.org/10.1016/j.compstruct.2017.08.088. PMID:WOS:000413321800004.

[34] Gray RW, Baird DG, Bohn JH. Effects of processing conditions on short TLCP fiber reinforced FDM parts. Rapid Prototyp J 1998;4(1):14−25. https://doi.org/10.1108/13552549810197514. PMID:WOS:000203405700002.

[35] Friedrich L, Collino R, Ray T, Begley M. Acoustic control of microstructures during direct ink writing of two-phase materials. Sens Actuator A-Phys. 2017;268:213−21. https://doi.org/10.1016/j.sna.2017.06.016. PMID:WOS:000418966700028.

[36] Deville S, Saiz E, Nalla RK, Tomsia AP. Freezing as a path to build complex composites. Science 2006;311(5760):515−8. https://doi.org/10.1126/science.1120937. Epub 2006/01/28, 16439659.

[37] Erb RM, Libanori R, Rothfuchs N, Studart AR. Composites reinforced in three dimensions by using low magnetic fields. Science 2012;335(6065):199−204. https://doi.org/10.1126/science.1210822. Epub 2012/01/17, 22246772.

[38] Kokkinis D, Schaffner M, Studart AR. Multimaterial magnetically assisted 3D printing of composite materials. Nat Commun 2015;6:10. https://doi.org/10.1038/ncomms9643. PMID:WOS:000364941500005.

[39] Zhu C, Liu TY, Qian F, Han TYJ, Duoss EB, Kuntz JD, Spadaccini CM, Worsley MA, Li Y. Supercapacitors based on three-dimensional hierarchical graphene aerogels with periodic macropores. Nano Lett 2016;16(6):3448−56. https://doi.org/10.1021/acs.nanolett.5b04965. PMID:WOS:000377642700006.

[40] Marascio MGM, Antons J, Pioletti DP, Bourban PE. 3D printing of polymers with hierarchical continuous porosity. Adv Mater Technol 2017;2(11):7. https://doi.org/10.1002/admt.201700145. PMID:WOS:000415150100002.

[41] Mac Murray BC, An X, Robinson SS, van Meerbeek IM, O'Brien KW, Zhao H, Shepherd RF. Poroelastic foams for Simple fabrication of complex soft robots. Adv Mater 2015;27(41):6334−40. https://doi.org/10.1002/adma.201503464.

[42] Mueller J, Raney JR, Shea K, Lewis JA. Architected lattices with high stiffness and toughness via multicore-shell 3D printing. Adv Mater 2018;30:1705001.

[43] Bartlett NW, Tolley MT, Overvelde JT, Weaver JC, Mosadegh B, Bertoldi K, Whitesides GM, Wood RJ. SOFT ROBOTICSA. 3D-printed, functionally graded soft robot powered by combustion. Science 2015;349(6244):161−5. https://doi.org/10.1126/science.aab0129. Epub 2015/07/15. PMID:26160940.

[44] Bartlett NW, Tolley MT, Overvelde JTB, Weaver JC, Mosadegh B, Bertoldi K, Whitesides GM, Wood RJ. A 3D-printed, functionally graded soft robot powered by combustion. Science 2015;349(6244):161−5. https://doi.org/10.1126/science.aab0129. PMID:WOS:000357664300036.

[45] Yap HK, Ng HY, Yeow CH. High-force soft printable pneumatics for soft robotic applications. Soft Robot 2016;3(3):144−58. https://doi.org/10.1089/soro.2016.0030. PMID:WOS:000384227200006.

[46] Cohrs NH, Petrou A, Loepfe M, Yliruka M, Schumacher CM, Kohll AX, Starck CT, Schmid Daners M, Meboldt M, Falk V, Stark WJ. A soft total artificial heart—first concept evaluation on a hybrid mock circulation. Artif Organs 2017;41(10):948−58. https://doi.org/10.1111/aor.12956.

[47] Buckberg G, Mahajan A, Saleh S, Hoffman JIE, Coghlan C. Structure and function relationships of the helical ventricular myocardial band. J Thorac Cardiovasc Surg 2008;136(3):578−589.e11. https://doi.org/10.1016/j.jtcvs.2007.10.088.

[48] Peele BN, Wallin TJ, Zhao HC, Shepherd RF. 3D printing antagonistic systems of artificial muscle using projection stereolithography. Bioinspiration Biomimetics 2015;10(5):8. https://doi.org/10.1088/1748-3190/10/5/055003. PMID:WOS:000363543700004.

[49] Ionov L. Biomimetic hydrogel-based actuating systems. Adv Funct Mater 2013;23(36):4555−70. https://doi.org/10.1002/adfm.201203692.

[50] Lee H, Xia C, Fang NX. First jump of microgel; actuation speed enhancement by elastic instability. Soft Matter 2010;6(18):4342−5. https://doi.org/10.1039/C0SM00092B.

[51] Sydney Gladman A, Matsumoto EA, Nuzzo RG, Mahadevan L, Lewis JA. Biomimetic 4D printing. Nat Mater 2016;15:413. 10.1038/nmat4544 https://http://www.nature.com/articles/nmat4544-supplementary-information.

[52] Wei HQ, Zhang QW, Yao YT, Liu LW, Liu YJ, Leng JS. Direct-write fabrication of 4D active shape-changing structures based on a shape memory polymer and its nanocomposite. ACS Appl Mater Interfaces 2017;9(1):876−83. https://doi.org/10.1021/acsami.6b12824. PMID:WOS:000392037400101.

[53] Ambulo CP, Burroughs JJ, Boothby JM, Kim H, Shankar MR, Ware TH. Four-dimensional printing of liquid crystal elastomers. ACS Appl Mater Interfaces 2017;9(42):37332−9. https://doi.org/10.1021/acsami.7b11851. PMID:WOS:000414115700092.

[54] Miriyev A, Stack K, Lipson H. Soft material for soft actuators. Nat Commun 2017; 8(1):596. https://doi.org/10.1038/s41467-017-00685-3.

[55] Ahn BY, Duoss EB, Motala MJ, Guo X, Park S, Xiong Y, Yoon J, Nuzzo RG, Rogers JA, Lewis JA. Omnidirectional printing of flexible, stretchable, and spanning silver microelectrodes. Science 2009;323:1590−3.

[56] Kim K, Zhu W, Qu X, Aaronson C, McCall WR, Chen SC, Sirbuly DJ. 3D optical printing of piezoelectric nanoparticle - polymer composite materials. ACS Nano 2014;8(10): 9799−806. https://doi.org/10.1021/nn503268f. PMID:WOS:000343952600012.

[57] Rossiter J, Walters P, Stoimenov B, editors. Printing 3D dielectric elastomer actuators for soft robotics. SPIE smart structures and materials + nondestructive evaluation and health monitoring. SPIE; 2009.

[58] Anderson IA, Gisby TA, McKay TG, O'Brien BM, Calius EP. Multi-functional dielectric elastomer artificial muscles for soft and smart machines. J Appl Phys 2012;112(4): 041101. https://doi.org/10.1063/1.4740023.

[59] Robinson SS, O'Brien KW, Zhao H, Peele BN, Larson CM, Mac Murray BC, Van Meerbeek IM, Dunham SN, Shepherd RF. Integrated soft sensors and elastomeric actuators for tactile machines with kinesthetic sense. Extreme Mech Lett 2015;5:47−53. https://doi.org/10.1016/j.eml.2015.09.005.

[60] 3D-printed ionic polymer-metal composite soft crawling robot. In: Carrico JD, Kim KJ, Leang KK, editors. 2017 IEEE International conference on robotics and automation (ICRA); May 29−June 3, 2017.

[61] Bertassoni LE, Cardoso JC, Manoharan V, Cristino AL, Bhise NS, Araujo WA, Zorlutuna P, Vrana NE, Ghaemmaghami AM, Dokmeci MR, Khademhosseini A. Direct-write bioprinting of cell-laden methacrylated gelatin hydrogels. Biofabrication 2014;6(2):024105. https://doi.org/10.1088/1758-5082/6/2/024105.

[62] Jia J, Richards DJ, Pollard S, Tan Y, Rodriguez J, Visconti RP, Trusk TC, Yost MJ, Yao H, Markwald RR, Mei Y. Engineering alginate as bioink for bioprinting. Acta Biomater 2014;10(10):4323−31. https://doi.org/10.1016/j.actbio.2014.06.034.

[63] Kolesky DB, Truby RL, Gladman AS, Busbee TA, Homan KA, Lewis JA. 3D bioprinting of vascularized, heterogeneous cell-laden tissue constructs. Adv Mater 2014;26(19):3124−30. https://doi.org/10.1002/adma.201305506.

[64] Williams SK, Touroo JS, Church KH, Hoying JB. Encapsulation of adipose stromal vascular fraction cells in alginate hydrogel spheroids using a direct-write three-dimensional printing system. Biores Open Access 2013;2(6):448−54. https://doi.org/10.1089/biores.2013.0046.

[65] Mironov V, Visconti RP, Kasyanov V, Forgacs G, Drake CJ, Markwald RR. Organ printing: tissue spheroids as building blocks. Biomaterials 2009;30(12):2164−74. https://doi.org/10.1016/j.biomaterials.2008.12.084.

[66] Kolesky DB, Homan KA, Skylar-Scott MA, Lewis JA. Three-dimensional bioprinting of thick vascularized tissues. Proc Natl Acad Sci USA 2016;113(12):3179−84. https://doi.org/10.1073/pnas.1521342113.

[67]   Homan KA, Kolesky DB, Skylar-Scott MA, Herrmann J, Obuobi H, Moisan A, Lewis JA. Bioprinting of 3D convoluted renal proximal tubules on perfusable chips. Sci Rep 2016;6:34845. https://doi.org/10.1038/srep34845.

[68]   Miller JS, Stevens KR, Yang MT, Baker BM, Nguyen D-HT, Cohen DM, Toro E, Chen AA, Galie PA, Yu X, Chaturvedi R, Bhatia SN, Chen CS. Rapid casting of patterned vascular networks for perfusable engineered three-dimensional tissues. Nat Mater 2012;11(9):768−74. https://doi.org/10.1038/nmat3357.

[69]   Ozbolat IT, Hospodiuk M. Current advances and future perspectives in extrusion-based bioprinting. Biomaterials 2016;76:321−43. https://doi.org/10.1016/j.biomaterials.2015.10.076.

[70]   Murphy SV, Atala A. 3D bioprinting of tissues and organs. Nat Biotechnol 2014;32(8):773−85. https://doi.org/10.1038/nbt.2958.

[71]   Duan B. State-of-the-Art review of 3D bioprinting for cardiovascular tissue engineering. Ann Biomed Eng 2017;45(1):195−209. https://doi.org/10.1007/s10439-016-1607-5.

[72]   Ji S, Guvendiren M. Recent advances in bioink design for 3D bioprinting of tissues and organs. Front Bioeng Biotechnol 2017;5:23. https://doi.org/10.3389/fbioe.2017.00023.

[73]   Trappmann B, Baker BM, Polacheck WJ, Choi CK, Burdick JA, Chen CS. Matrix degradability controls multicellularity of 3D cell migration. Nat Commun 2017;8(1). https://doi.org/10.1038/s41467-017-00418-6.

[74]   Gudapati H, Dey M, Ozbolat I. A comprehensive review on droplet-based bioprinting: past, present and future. Biomaterials 2016;102:20−42. https://doi.org/10.1016/j.biomaterials.2016.06.012.

[75]   Hölzl K, Lin S, Tytgat L, Van Vlierberghe S, Gu L, Ovsianikov A. Bioink properties before, during and after 3D bioprinting. Biofabrication 2016;8(3):032002. https://doi.org/10.1088/1758-5090/8/3/032002.

[76]   Malda J, Visser J, Melchels FP, Jüngst T, Hennink WE, Dhert WJA, Groll J, Hutmacher DW. 25th anniversary article: engineering hydrogels for biofabrication. Adv Mater 2013;25(36):5011−28. https://doi.org/10.1002/adma.201302042.

[77]   Liu J, Zheng H, Poh PSP, Machens H-G, Schilling AF. Hydrogels for engineering of perfusable vascular networks. Int J Mol Sci 2015;16(7):15997−6016. https://doi.org/10.3390/ijms160715997.

[78]   Huang G, Wang L, Wang S, Han Y, Wu J, Zhang Q, Xu F, Lu TJ. Engineering three-dimensional cell mechanical microenvironment with hydrogels. Biofabrication 2012;4(4):042001. https://doi.org/10.1088/1758-5082/4/4/042001.

[79]   Skardal A, Atala A. Biomaterials for integration with 3-D bioprinting. Ann Biomed Eng 2015;43(3):730−46. https://doi.org/10.1007/s10439-014-1207-1.

[80]   Bryant SJ, Bender RJ, Durand KL, Anseth KS. Encapsulating chondrocytes in degrading PEG hydrogels with high modulus: engineering gel structural changes to facilitate cartilaginous tissue production. Biotechnol Bioeng 2004;86(7):747−55. https://doi.org/10.1002/bit.20160.

[81]   Nguyen DG, Funk J, Robbins JB, Crogan-Grundy C, Presnell SC, Singer T, Roth AB. Bioprinted 3D primary liver tissues allow assessment of organ-level response to clinical drug induced toxicity in vitro. PLoS One 2016;11(7):e0158674. https://doi.org/10.1371/journal.pone.0158674.

[82]   Pati F, Jang J, Ha D-H, Won Kim S, Rhie J-W, Shim J-H, Kim D-H, Cho D-W. Printing three-dimensional tissue analogues with decellularized extracellular matrix bioink. Nat Commun 2014;5:3935. https://doi.org/10.1038/ncomms4935.

[83] Jang J, Park H-J, Kim S-W, Kim H, Park JY, Na SJ, Kim HJ, Park MN, Choi SH, Park SH, Kim SW, Kwon S-M, Kim P-J, Cho D-W. 3D printed complex tissue construct using stem cell-laden decellularized extracellular matrix bioinks for cardiac repair. Biomaterials 2017;112:264−74. https://doi.org/10.1016/j.biomaterials.2016.10.026.

[84] Jakus AE, Rutz AL, Shah RN. Advancing the field of 3D biomaterial printing. Biomed Mater 2016;11(1):014102. https://doi.org/10.1088/1748-6041/11/1/014102.

[85] Wang X, Yan Y, Pan Y, Xiong Z, Liu H, Cheng J, Liu F, Lin F, Wu R, Zhang R, Lu Q. Generation of three-dimensional hepatocyte/gelatin structures with rapid prototyping system. Tissue Eng 2006;12(1):83−90. https://doi.org/10.1089/ten.2006.12.83.

[86] Maher PS, Keatch RP, Donnelly K, Mackay RE, Paxton JZ. Construction of 3D biological matrices using rapid prototyping technology. Rapid Prototyp J 2009;15(3): 204−10. https://doi.org/10.1108/13552540910960307.

[87] Smith CM, Christian JJ, Warren WL, Williams SK. Characterizing environmental factors that impact the viability of tissue-engineered constructs fabricated by a direct-write bioassembly tool. Tissue Eng 2007;13(2). https://doi.org/10.1089/ten.2007.13.ft-338. 070108042223001.

[88] Lee W, Lee V, Polio S, Keegan P, Lee J-H, Fischer K, Park J-K, Yoo S-S. On-demand three-dimensional freeform fabrication of multi-layered hydrogel scaffold with fluidic channels. Biotechnol Bioeng 2010. https://doi.org/10.1002/bit.22613.

[89] Park SA, Lee SH, Kim W. Fabrication of hydrogel scaffolds using rapid prototyping for soft tissue engineering. Macromol Res 2011;19(7):694−8. https://doi.org/10.1007/s13233-011-0708-0.

[90] Gyles DA, Castro LD, Silva JOC, Ribeiro-Costa RM. A review of the designs and prominent biomedical advances of natural and synthetic hydrogel formulations. Eur Polym J 2017;88:373−92. https://doi.org/10.1016/j.eurpolymj.2017.01.027.

[91] Duan B, Hockaday LA, Kang KH, Butcher JT. 3D bioprinting of heterogeneous aortic valve conduits with alginate/gelatin hydrogels. J Biomed Mater Res 2013;101(5): 1255−64. https://doi.org/10.1002/jbm.a.34420.

[92] Khalil S, Sun W. Bioprinting endothelial cells with alginate for 3D tissue constructs. J Biomech Eng 2009;131(11):111002. https://doi.org/10.1115/1.3128729.

[93] Moon S, Hasan SK, Song YS, Xu F, Keles HO, Manzur F, Mikkilineni S, Hong JW, Nagatomi J, Haeggstrom E, Khademhosseini A, Demirci U. Layer by layer three-dimensional tissue epitaxy by cell-laden hydrogel droplets. Tissue Eng Part C Methods 2010;16(1):157−66. https://doi.org/10.1089/ten.TEC.2009.0179.

[94] Hockaday LA, Kang KH, Colangelo NW, Cheung PYC, Duan B, Malone E, Wu J, Girardi LN, Bonassar LJ, Lipson H, Chu CC, Butcher JT. Rapid 3D printing of anatomically accurate and mechanically heterogeneous aortic valve hydrogel scaffolds. Biofabrication 2012;4(3):035005. https://doi.org/10.1088/1758-5082/4/3/035005.

[95] Lee VK, Kim DY, Ngo H, Lee Y, Seo L, Yoo S-S, Vincent PA, Dai G. Creating perfused functional vascular channels using 3D bio-printing technology. Biomaterials 2014;35(28):8092−102. https://doi.org/10.1016/j.biomaterials.2014.05.083.

[96] Chen Y-C, Lin R-Z, Qi H, Yang Y, Bae H, Melero-Martin JM, Khademhosseini A. Functional human vascular network generated in photocrosslinkable gelatin methacrylate hydrogels. Adv Funct Mater 2012;22(10):2027−39. https://doi.org/10.1002/adfm.201101662.

[97] Loessner D, Meinert C, Kaemmerer E, Martine LC, Yue K, Levett PA, Klein TJ, Melchels FPW, Khademhosseini A, Hutmacher DW. Functionalization, preparation and use of cell-laden gelatin methacryloyl-based hydrogels as modular tissue culture platforms. Nat Protoc 2016;11(4):727−46. https://doi.org/10.1038/nprot.2016.037.

[98] Duan B, Kapetanovic E, Hockaday LA, Butcher JT. Three-dimensional printed trileaflet valve conduits using biological hydrogels and human valve interstitial cells. Acta Biomater 2014;10(5):1836−46. https://doi.org/10.1016/j.actbio.2013.12.005.

[99] Gaetani R, Feyen DAM, Verhage V, Slaats R, Messina E, Christman KL, Giacomello A, Doevendans PAFM, Sluijter JPG. Epicardial application of cardiac progenitor cells in a 3D-printed gelatin/hyaluronic acid patch preserves cardiac function after myocardial infarction. Biomaterials 2015;61:339−48. https://doi.org/10.1016/j.biomaterials.2015.05.005.

[100] Schuurman W, Levett PA, Pot MW, van Weeren PR, Dhert WJA, Hutmacher DW, Melchels FPW, Klein TJ, Malda J. Gelatin-methacrylamide hydrogels as potential biomaterials for fabrication of tissue-engineered cartilage constructs. Macromol Biosci 2013;13(5):551−61. https://doi.org/10.1002/mabi.201200471.

[101] Skardal A, Zhang J, McCoard L, Xu X, Oottamasathien S, Prestwich GD. Photocrosslinkable hyaluronan-gelatin hydrogels for two-step bioprinting. Tissue Eng Part A 2010;16(8):2675−85. https://doi.org/10.1089/ten.tea.2009.0798.

[102] Nichol JW, Koshy ST, Bae H, Hwang CM, Yamanlar S, Khademhosseini A. Cell-laden microengineered gelatin methacrylate hydrogels. Biomaterials 2010;31(21):5536−44. https://doi.org/10.1016/j.biomaterials.2010.03.064.

[103] Lim KS, Schon BS, Mekhileri NV, Brown GCJ, Chia CM, Prabakar S, Hooper GJ, Woodfield TBF. New visible-light photoinitiating system for improved print fidelity in gelatin-based bioinks. ACS Biomater Sci Eng 2016;2(10):1752−62. https://doi.org/10.1021/acsbiomaterials.6b00149.

[104] Skardal A, Mack D, Kapetanovic E, Atala A, Jackson JD, Yoo J, Soker S. Bioprinted amniotic fluid-derived stem cells accelerate healing of large skin wounds. Stem Cells Transl Med 2012;1(11):792−802. https://doi.org/10.5966/sctm.2012-0088.

[105] Ahmed TAE, Dare EV, Hincke M. Fibrin: a versatile scaffold for tissue engineering applications. Tissue Eng Part B Rev 2008;14(2):199−215. https://doi.org/10.1089/ten.teb.2007.0435.

[106] Duan B, Hockaday LA, Kapetanovic E, Kang KH, Butcher JT. Stiffness and adhesivity control aortic valve interstitial cell behavior within hyaluronic acid based hydrogels. Acta Biomater 2013;9(8):7640−50. https://doi.org/10.1016/j.actbio.2013.04.050.

[107] Rodell CB, MacArthur JW, Dorsey SM, Wade RJ, Wang LL, Woo YJ, Burdick JA. Shear-thinning supramolecular hydrogels with secondary autonomous covalent crosslinking to modulate viscoelastic properties. Adv Funct Mater 2015;25(4):636−44. https://doi.org/10.1002/adfm.201403550.

[108] Burdick JA, Prestwich GD. Hyaluronic acid hydrogels for biomedical applications. Adv Mater 2011;23(12):H41−56. https://doi.org/10.1002/adma.201003963.

[109] Highley CB, Rodell CB, Burdick JA. Direct 3D printing of shear-thinning hydrogels into self-healing hydrogels. Adv Mater 2015;27(34):5075−9. https://doi.org/10.1002/adma.201501234.

[110] Ouyang L, Highley CB, Rodell CB, Sun W, Burdick JA. 3D printing of shear-thinning hyaluronic acid hydrogels with secondary cross-linking. ACS Biomater Sci Eng 2016;2(10):1743−51. https://doi.org/10.1021/acsbiomaterials.6b00158.

[111] Kang H-W, Lee SJ, Ko IK, Kengla C, Yoo JJ, Atala AA. 3D bioprinting system to produce human-scale tissue constructs with structural integrity. Nat Biotechnol 2016; 34(3):312−9. https://doi.org/10.1038/nbt.3413.

[112] Paulsen SJ, Miller JS. Tissue vascularization through 3D printing: will technology bring us flow? Dev Dyn 2015;244(5):629−40. https://doi.org/10.1002/dvdy.24254.

[113] Kucukgul C, Ozler SB, Inci I, Karakas E, Irmak S, Gozuacik D, Taralp A, Koc B. 3D bioprinting of biomimetic aortic vascular constructs with self-supporting cells. Biotechnol Bioeng 2015;112(4):811−21. https://doi.org/10.1002/bit.25493.

[114] Poldervaart MT, Gremmels H, van Deventer K, Fledderus JO, Oner FC, Verhaar MC, Dhert WJA, Alblas J. Prolonged presence of VEGF promotes vascularization in 3D bioprinted scaffolds with defined architecture. J Control Release 2014;184:58−66. https://doi.org/10.1016/j.jconrel.2014.04.007.

[115] Norotte C, Marga FS, Niklason LE, Forgacs G. Scaffold-free vascular tissue engineering using bioprinting. Biomaterials 2009;30(30):5910−7. https://doi.org/10.1016/j.biomaterials.2009.06.034.

[116] Mirabella T, MacArthur JW, Cheng D, Ozaki CK, Woo YJ, Yang MT, Chen CS. 3D-printed vascular networks direct therapeutic angiogenesis in ischaemia. Nat Biomed Eng 2017;1(6):0083. https://doi.org/10.1038/s41551-017-0083.

[117] Gao Q, He Y, Fu J-Z, Liu A, Ma L. Coaxial nozzle-assisted 3D bioprinting with built-in microchannels for nutrients delivery. Biomaterials 2015;61:203−15. https://doi.org/10.1016/j.biomaterials.2015.05.031.

[118] Jia W, Gungor-Ozkerim PS, Zhang YS, Yue K, Zhu K, Liu W, Pi Q, Byambaa B, Dokmeci MR, Shin SR, Khademhosseini A. Direct 3D bioprinting of perfusable vascular constructs using a blend bioink. Biomaterials 2016;106:58−68. https://doi.org/10.1016/j.biomaterials.2016.07.038.

Applications of 3D Printing

Simon Dunham[1,2], **Bobak Mosadegh**[1,2], **Eva A. Romito**[2], **Mohamed Zgaren**[2]

Department of Radiology, Weill Cornell Medicine, New York, NY, United States[1]; Dalio Institute of Cardiovascular Imaging, NewYork-Presbyterian Hospital, New York, NY, United States[2]

3D PRINTING FOR THE CONSUMER MARKET

3D printing has been hyped in popular culture since the early 2010s as a method that would revolutionize the consumer market, by enabling the on-demand fabrication of consumer goods in the comfort of a person's home [1,2]. This vision is inspired by technology that was popularized in science fiction films, such as the Replicator in *Star Trek*. Although 3D printing is not expected to realize this vision in the near future, it has already made a significant impact in several fields, including the consumer market [3].

Currently, the consumer market for 3D printing can be broken down into three categories: (1) *Purchasing 3D printers for at-home use.* Although industrial 3D printers still make up a vast majority of the market, the use of desktop 3D printers is growing as this technology becomes more accessible with smaller form factors, lower costs, and increased availability of materials. There are over 60 consumer style 3D printers available on the market [4], ranging from a couple of hundred to several thousands of dollars in price (Fig. 4.1). Originally hobbyists and small businesses were the primary consumers of desktop 3D printers, but this market has grown to include educational institutions [5], ranging from k-12 schools to large universities. Furthermore, there is a growing market for purchasing designs to be printed on personal 3D printers. There are now several online websites (e.g., Thingiverse, Pinshape, Sketchfab) that host thousands of printable files that can be downloaded for a fee or for free. (2) *Ordering parts designed by the consumer to be 3D printed and shipped from a "fabshop."* The primary skill needed for 3D printing is to use computer-aided design (CAD) software to generate the shapes needed for printing. As these softwares become more user-friendly and lower in cost, more and more users are able to generate these complex drawings and then have the drawings fabricated by a fabshop and delivered to their home or office. Shapeways and Sculpteo are companies devoted to 3D printing and allow users to upload printable files that can be printed in a variety of materials, ranging from plastics and metals to ceramics. (3) *Purchasing 3D printed products sold online by designers/artisans.* A major feature of 3D printing is the ability to

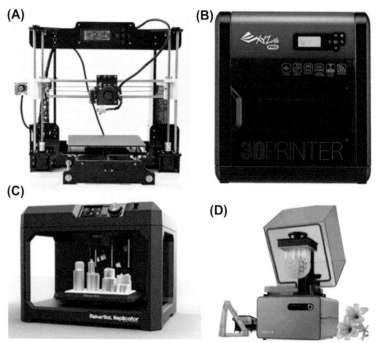

FIGURE 4.1

Images of available consumer-style 3D printers. (A) 3D printed SRB driver face (printed in color, sold for $4.93). (B) da Vinci Pro 3D Printer (sold for $667.92). (C) fifth generation MakerBot Replicator (sold for $2899.00). (D) Formlab's Form 1$^+$ stereolithography (sold for $3299.00).

Reproduced with permission from Langnau L. Testing consumer 3D printers 2015. Available from: http://www.makepartsfast.com/testing-consumer-3d-printers/.

fabricate objects with complex designs at a small scale, thus making 3D printing an extremely useful tool for product designers and artisans to sell unique products that are likely to be sold in small quantities. Shapeways, for example, has built-in "shops" on their websites, allowing individuals to sell their designs and advertise their brand. Since Shapeways will 3D print the product on demand, there is no need for the seller to have any inventory or equipment, thus eliminating any upfront investment (Fig. 4.2). An alternative platform is Etsy, which is a website that showcases handmade, vintage, and/or uniquely manufactured items that are sold peer to peer; this popular website also showcases 3D printed parts, since they are often niche items.

There are key limitations that have prevented the growth of the consumer market for 3D printing [6]. The first limitation is the availability of various materials. Although 3D printers have been made for a variety of materials, these have largely been sold for industrial and research institutions, due to their high cost. Another limitation is the need for postprocessing steps, since many 3D printers

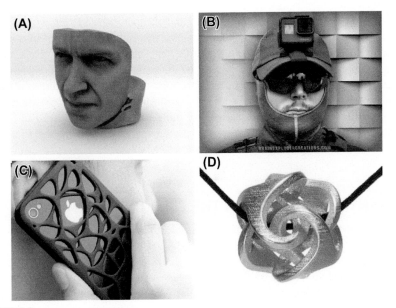

FIGURE 4.2

Images of available products on Shapeways website. (A) 3D printed SRB driver face (printed in color, sold for $4.93). (B) 3D printed hat mount for the GoPro HERO 5 (sold for $49.99). (C) 3D printed double layer iPhone case (sold for $23.49). (D) 3D printed Ora pendant in gold steel (sold for $19.00).

Figure reproduced from Shapeways.com.

require removal of support material, washing out of uncured resin, and/or smoothing of surfaces. Depending on the geometry of the part, these steps can be very time-consuming, preventing 3D printing to be the print-and-play process expected by consumers. A third limitation is the need for preprocessing, to ensure the file will produce the intended part correctly. This process is actually fairly mature as most 3D printers come with their own software to convert ".stl" files into g-code with little knowledge of 3D printing required. However, more complex objects that have composite materials or gradients of materials still require manual selection steps that are cumbersome for most consumer use. Therefore, consumer 3D printers are currently sold to fabricate plastic parts (e.g., acrylonitrile butadiene styrene (ABS), polylactic acid (PLA)) since the technology behind these printers is fairly inexpensive and the postprocessing steps are minimal. Stereolithography (SLA) and digital light processing printers have become more popular, due to their increased resolution and speed, but the postprocessing steps have limited their use to more specialized users, such as hobbyists.

A major factor that will significantly influence the use of 3D printing for consumer products is the increase in ease of printing and reduction in cost of distribution. With the ever-progressing efficiency of logistics, coordinating delivery of products to consumers is becoming faster and cheaper, currently allowing certain items to be ordered and received on the same day; this capability renders on-demand fabrication in the home of the consumer to be less and less desirable. Thus, many consumer reports predict that production of 3D printers for retail purchases will decrease [7], and consumers will utilize fabshops to order their customized parts. These types of orders will increase as prices drop and users become more and more proficient at 3D CAD. Although 3D printing is seen to be most fruitful for industrial applications, the pendulum could swing the other way in the future as printers become cheaper, more compact, and easier to use.

3D PRINTING EDIBLES AND THE FOOD INDUSTRY

The potential to create novel and customizable food products has driven interest in 3D printed food products; however, this has yet to become a practical process with widespread utility. The goals in personalized nutrition, use of alternative food sources, and alleviation of dietary allergens are also strong drivers for the pragmatic development and research in this niche field. However, it is the on-demand production and trends in modernist cuisine that have historically garnered more attention and are the regnant applications of this technology.

Similar to other applications of 3D printing, or additive manufacturing, the primary methods by which edible constructs are achieved are extrusion and inkjet processes [8,9]. Three main identifiable strategies with distinct goals/motivation in the pursuit of 3D printed edibles are bio-driven, bottom-up, and top-down [10,11]. Briefly, research initiatives that fall into the bio-driven category follow a school of thought akin to researchers pursuing tissue engineering (TE) with a differing final outcome. The pursuit of highly complex tissue structures for bio-driven food research primarily targets muscle biomimicry for the establishment of batch-produced meats that can be incubated for consumption. Bottom-up initiatives are more concerned with problems surrounding a limited food supply with a growing population. Using alternative edible materials, this area of research seeks to re-create familiar foods by altering proportions and patterns of limited components. Top-down strategies involve the use of common ingredients (primarily those in the form of liquids or pastes) and incorporating additives to create customizable foodstuffs. The top-down approach to research in this field is presently the most advanced: early adopters in the food industry have already created products aimed at consumers from the food industry.

Methods for 3D printing of edible structures vary depending on the materials available for printing and the final requirements for content, cooking, texture, and flavor. The use of binder jet printing has been aggressively pursued for avant-garde textural applications by chefs and companies such as Sugar Lab, Inc. and 3D Systems. This technology uses an edible powder substrate, and ejects a binding liquid in the desired pattern. Alternating layers of powder and binding liquid eventually form a solid shape that can be consumed as intricate sugar sculptures, or further processed before consumption. The use of fused deposition modeling is the most common form of printing method in the current climate of printed foods favoring liquids, semisolid pastes, and shear-thinning components [12–14]. This type of extrusion-based method can be further broken down into different categories, namely extrusion cooking and extrusion printing. The cooking addendum clarifies the purpose and output of this technology as it seeks to be used at a larger commercial level. The input for extrusion cooking is raw or preground components that can be heated/cooled and shaped at the extruder, resulting in shelf-stable or fully cooked products (e.g., pasta). An example of extrusion-based food printing of a similarly familiar item can be found in the collaboration of Barilla with the Netherlands Organization for Applied Scientific Research (TNO) to create intricate 3D printed shapes of pasta done at a smaller batch-scale, with more customizable features (Fig. 4.3).

Extrusion food printing has also been highly popular in the creation of intricate shapes with semisolid, liquid, and foam materials such as chocolate, dough, processed vegetables/meats, and meringues [15–18]. The aesthetics, incorporation of mixed components for gastronomically interesting textures, and flavor profiles make this an attractive and creative approach for modernist cooking presentations. Companies such as ChocEdge, TNO, Natural Machines, NuFood, and Dovetail have produced several types of products (e.g., chocolates, jellies, cookies, and

FIGURE 4.3

3D printed edibles and printers currently in market. Top left, Netherlands Organization for Applied Scientific Research (TNO) printer. Center, Barilla collaboration with TNO. Right, Nufood Inc. printer. Bottom left, Chefjet by 3D Systems. Center, Sugar creations from the culinary lab by 3D Systems. Right, Discov3ry ultimaker, by makerlabs.

high nutrition pastes). The customization of food at a batch scale allows for enhanced aesthetics and taste, and the opportunity for health-conscious applications. The potential for control over all the components of a food product can be immensely beneficial in terms of nutrition, food allergen control, oral drug delivery, and textures for those suffering from dysphagia (difficulty swallowing) [19]. Creating nutritionally targeted edibles are already somewhat popular with the spike in health trends and nutritional supplements, but a more advanced regimented approach that is customized to the individual could be on the horizon. Parameters for an attractive edible material that can carry this type of supplement involve the characterization and tailoring of the material to ensure compatibility with current printing methods. An example was demonstrated by Derossi et al., who investigated specific printing variables related to the production of a nutritionally balanced fruit snack for children of a specific age group [9]. Creating nutritionally rich foods for people suffering from dysphagia has also been a compelling use of 3D food printing, and is represented in the efforts put forth by Dovetail and other groups, utilizing bioprinters with preprocessed components in paste form [15,16]. In the vein of treating disease, the possibility of pharmacological additives in personalized printed food is also an exciting future venture. Already, 3D printed tablets are being explored with multiple drugs as potential candidates for oral delivery [20−22]. While all these applications are not yet pragmatic uses of 3D printing technologies, the promise of this young field continues to be an appetizing endeavor.

MEDICAL

Because of the wide array of applications in healthcare, medical 3D printing calls upon all of the methods and materials for 3D printing discussed in Chapters 2 and 3 to allow the creation of custom-built medical models, devices, drugs, and training tools. The remaining chapters of this book provide much greater detail about the range of applications to consider, but we will provide a brief description here. An example of the unique potential and challenges of medical 3D printing is highlighted by the FDA's approval of a 3D printed drug product in August 2015 [23]. This shows the level of interest in the field, but also the unique regulatory hurdles that apply here. One aspect that is driving interest in medical 3D printing is the variability of the size and shape of the human body. There is a resulting demand for custom-made and dedicated medical devices. Today, 3D printing technology is used in a huge range of medical applications such as dentistry [24], drug and pharmaceutical fabrication [25], anatomical models [26], hearing aids [27], surgical instruments [28], TE [29], medical training and education [30] (Fig. 4.4).

The dental industry in particular is one of the more mature markets, largely due to the importance of precise fit in creating crowns, caps, and other dental prosthetics. In addition, many of the casting and molding processes used to convert dental models, made from nonimplantable materials, into dental implants have been a

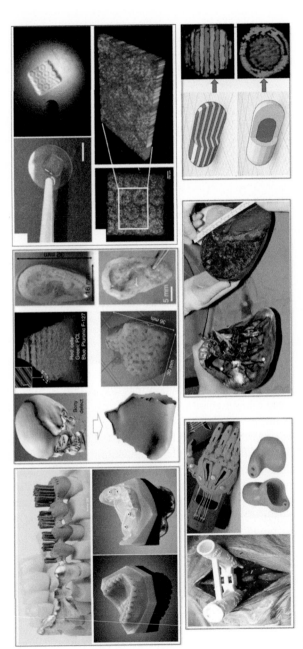

FIGURE 4.4

Applications of 3D printing in medicine. Examples of 3D printed products (*top left to bottom right*); dental products, bioprinted tissues in human geometries, bioprinted tissue constructs, hearing aids, presurgical planning models, and pharmaceuticals.

Reproduced with permission from Liaw CY, Guvendiren M. Current and emerging applications of 3D printing in medicine. Biofabrication 2017;9:024102.

Table 4.1 Challenges and Future Research

Field	Challenges
Dentistry	Enhancement of fabricated prototype quality by introducing new techniques and avoiding thermal distortion. Reducing the fabrication process time by gathering layers in one step.
Drug manufacturing	Improve drug loading efficacy. Reduce postfabrication process.
Medical devices	Reduce time and cost with quality improvement. Wider range of target medical application. Online service for specific applications.

part of the field for decades, thus making the acceptance of 3D printing easier. 3D printed models are used for restorations, surgical guides, dental models, and orthodontic tools [1].

In the medical field, regulations have limited the applications of 3D printing; however, there are still a variety of devices that have benefited. 3D printed models for visualization and surgical planning, custom surgical guides, and other surgical tools are becoming more common. In limited cases, 3D printed implants have also found use [1].

Despite the growth of 3D printing in the healthcare field, there are various challenges for this technology in the near future, especially regarding regulation concerns and safety rules. In 2016, the FDA introduced a technical guidance for the 3D printing industry—Technical Considerations for Additive Manufactured Devices [31]. The document highlighted two main progress categories: design and manufacturing, and testing of the devices. Most notably, FDA approvals for 3D printed devices must consider not only the final device (which still must meet the same rigorous standards for performance and biocompatibility), but also the process used to determine the device designs, the limits and margins of possible designs, and the performance of the designs most likely to undergo failure. In other words, device manufacturers must prove not only the performance of their design, but the performance of all possible designs. Table 4.1 summarizes the essential challenges that 3D printing will face in the future in the medical field.

INDUSTRIAL

For several decades, 3D printing has been employed in the preproduction, prototyping, and post-fabrication phases [32]. In the automotive and aerospace industries, the use of rapidly available 3D printed components as part of the design process is commonplace. This allows designers to iterate without waiting for complex tooling or other manufacturing processes to take place. More recently, these sectors have begun to explore the use of directly printed 3D parts for nonfunctional components. For automobiles, this includes components such as dashboard assemblies or hood ornaments [33]. 3D printing is especially useful for producing parts for old or rare cars or to allow consumers to have truly customized designs. 3D printing is also widely used by the aerospace industry due to the significant advantages in cost, accuracy, and rapidity. Aside from components, which can be used in the design, testing, and education process [34], many final parts are now 3D printed. Examples include seat frames and arm rest caps. Today, the A350 XWB aircraft has over a 1000 3D printed components [35]. In addition, for space travel, where almost all parts are custom-made, 3D printing produces many critical components [35]. While these applications have already had a significant impact, they are still relatively niche. The potential for further expansion in these sectors will likely need to be driven by new products, which feature design uniqueness and/or customizability as a selling point [33]. A variety of products are made using some form of additive manufacturing. In the construction industry, key advances are driven by the development of faster and larger area printers and new materials, especially concrete and other building materials [36]. Most notably, recently, Apis Cor, a startup, was able to 3D print an entire 400 sq. ft. home in 24 h [37]. The company cited the cost as roughly $10,000.

3D printing production in the industrial field is a promising technique that could bring huge benefits to manufacturing facilities. 3D printing currently has a significant impact on the industrial sector; however, several key advances have the potential to dramatically expand its utility:

- Printing multiple layers of material at the same time.
- Printing active devices like batteries and embedded machines.
- Printing constructions, like houses and building prototypes.
- Printing complex items and products for vehicles and trucks.
- Printing in harsh and tough environments, such as under water and in space.

3D BIOPRINTING

Advances in bioprinting/additive manufacturing technologies have allowed for the creation of increasingly complex structures and elegant bioink systems in order to support the study of various tissues. The advent of improved spatial resolution (down to 2 μm) and biocompatible materials has continually pushed forward the exploration of bioprinting in a laboratory setting. The three primary methods of bioprinting—inkjet, laser-assisted, and extrusion bioprinting—are still in their infancy as the ideal future for complete organ replacement and advanced regenerative medicine (RM). However, a considerable amount of success has been achieved in laboratory settings for applications in TE, RM, and in vitro drug screening [38–41].

Additive manufacturing of living cells, or bioprinting, works in concert with materials suitable for the support of complex structures mimicking natural environments. Natural and synthetic polymers used as primary components and supporting substrates have been shown to have substantial impact on cells that are contained within the created constructs. Valuable characteristics identified in the biocompatible material of choice vary amongst bioprinting methods, due to limitations in printing fidelity, but include: viscosity, shear-thinning, viscoelasticity, gelation kinetics, swelling, and biodegradation [38,42]. Additionally, the incorporation of bioactive moieties for cell adhesion, proliferation, and in some cases differentiation can dictate the bioink/printing substrate, as well as the most appropriate printing method. Cell viability is an inherent influencing factor in the success of bioprinting initiatives, as the direct printing of cells has distinct advantages in rapidly increasing the time to create a construct and improving the complexity of such bioprinted constructs. The direct printing of cells presents several challenges, such as the control of pH, shear stress, temperature, and osmotic pressure of potential bioink materials. The use of decellularized extracellular matrix and other naturally derived polymers such as alginates, chitosan derivatives, and hyaluronan systems are attractive for direct cell printing via inkjet deposition. SLA, laser-assisted methods, and extrusion-based methods have found more success with synthetic hydrogel compositions of pluronics, poly(ethylene glycol) systems, and interpenetrating polymer networks. Inkjet bioprinting presents the most affordable option of the present technologies and utilizes two basic methods for ink ejection: piezoelectric actuators that expel droplets when actuated or a heating element that creates a vapor bubble that expands and eventually ejects a droplet of bioink on to a substrate. SLA and extrusion-based methods that utilize photocross-linking can result in damage to encapsulated cells, and are thereby more often used in preparations where cells are seeded post printing. Extrusion-based methods normally utilize a heated nozzle tip which is used to smoothly melt and deposit polymers, but in bioprinting applications, shear-thinning behavior of bioinks allows for direct extrusion by a differential pressure gradient. Bioinks can be extruded onto a surface and are often further cross-linked by photopolymerization methods or by deposition on to a substrate that facilitates cross-linking (i.e., sodium alginate and calcium chloride). Laser-assisted bioprinting (LAB) methods have become popular due to their fine resolution (1−125 μm range), but present a heavy cost limitation due to their components. LAB systems require a direct-write method, whereby a pulsed laser source directly prints onto a "ribbon" of a nonabsorbing substrate coated in a thin layer of laser absorbing metal. Cells are contained in a liquid medium and are coated atop the metal layer. Here, deposition of droplets occurs when the laser writes on the ribbon, producing vaporization of the metal layer that then forces a droplet of cell containing ink to deposit onto a surface. Other methods that make use of powder binding and selective laser sintering are not compatible with direct cell printing.

Recent trends in the creation of structures with heterogeneous architecture and increasing complexity have moved toward multimaterial deposition [43,44], and the use of multiple cell types (as primary and support cells). Naturally, this presents new capabilities in the mimicry of native tissue architectures, with attempts at complex vascular networks, organoid structures, and coculture systems for toxicity and drug screening. Primary targets of TE have been found in skin, bone/cartilage, neuronal, and cardiac applications. The realization of integument in the appropriate functionalization of layers has had significant advances in composition and personalization, but has yet to produce results that are a practical answer to clinical needs of large areas of coverage for wounds and burns with a reasonable production timeline. Bone and cartilaginous tissue research has seen developments in fabrication branching into inkjet printing [45,46], as well as new strategies with novel bioinks used in conjunction with extrusion-based and LAB methods [47–50]. The appearance of in situ approaches for bone TE presents a new perspective and an era in its application. Neuronal TE/RM targets in 3D printing have similarly relied on extrusion as the main mode of fabrication, but new additives in bioinks, such as graphene and other polysaccharides have demonstrated success in the differentiation of neuronal tissue [51]. Continuing the theme of novel bioink usage finds cardiac applications of note, with extrusion of scaffolds that allow for complex structures and incorporation of electrical stimulation during incubation periods—effectively trying to re-create the structure and function of working cardiomyocytes for RM [52]. Additionally, patient-specific harvesting of stem cells and differentiation into cardiomyocytes have shaped changes to the structure and development of bioinks with specific bioactive payloads [53].

As bioprinting techniques refined with exciting prospects such as in situ and direct 3D printing with new hydrogels/bioinks, pluronics, and interpenetrating polymer networks are realized, the true value of not only patient-specific, but application-specific methods will become a more tangible possibility. The potential contributions to basic scientific understanding of personalized guest-host material interactions and drug screening are other growing regions of interest that bioprinting can be expected to significantly enrich [54–56].

POTENTIAL OF 3D PRINTING FOR CLOTHING AND ACCESSORIES

Because of the highly personalized nature of style and design, the idea of custom 3D printed clothing has been an area of interest since the advent of the technology. There are several aspects of 3D printing that are intrinsically desirable for fashion, clothing, and accessories. For customers, 3D printed clothing and accessories can be much more customizable than clothing produced through traditional manufacturing. This means 3D printing provides the ability to adjust the size and shape of clothing to the customer's body, and even allows for the possibility of 3D scanning an individual and producing custom bespoke clothing. In addition, there are more ways that customers could adjust designs to create features unique to them. Aside from aspects of design that are desirable, 3D printing provides the possibility of disrupting traditional supply chain models allowing for the number and type of clothing articles to be better suited to the customer demand, allowing vendors to avoid clearance and overstocking. Finally, the potential for individuals and hobbyists to be able to design and fabricate their own clothing provides the potential for a fashion revolution by lowering the barrier to participate in the design space [57–59].

For all these reasons, 3D printing has been thought to make a significant impact on the clothing and accessory industry for some time. However, to date, most demonstrations are limited to concept pieces, high fashion, or very limited release products. A variety of perspectives have attempted to understand the reasons for the gap between the potential of the technology and its current acceptance [60,61]. The most obvious causes are still related to the technological limitations of 3D printing. The cost of 3D printed parts and time associated with production are still relatively high compared to traditional manufacturing methods, when produced at scale. Furthermore, materials with the flexibility and durability required to provide robust articles are still challenging to produce, with only a couple commercial technologies offering flexible materials. It is worth noting that a variety of very recent technologies provide promise to overcome these challenges. Most notably, Stratasys flexible Tango materials provide the ability to produce articles with spatially varying mechanical properties. Fig. 4.5 shows "Caress of the Gaze" by Behnaz Farahi, a dynamic article of clothing with fiber that has spatially varying mechanics on a flexible backbone, which can be actuated in response to the gaze of an observer based on a camera and series of actuators [62]. Notably, Carbon 3D's thermoplastic polyurethane (TPU) materials provide flexibility and their proprietary printing technology allows for very rapid production of parts. They have teamed with Adidas to produce a limited release of the Futurecraft 3D, shoes with a 3D printed midsole [63].

Studies that attempt to understand the factors most strongly influencing potential customer acceptance of 3D printed clothing emphasized a variety of factors. Consumers' attitude about 3D printing was a strong determinant of their intention to utilize the product, and this was influenced by the garment's attractiveness, usefulness, and performance, in that order [64].

3D PRINTED TEXTILES

While many applications are currently limited by the cost and materials associated with printers, there is no doubt that 3D printed textiles will provide new options for fashion. In particular, the ability to tailor the pattern and mechanics of textiles provides for many new possibilities in terms of textile mechanics. A number of designers are considering the ways these patterns can affect the properties of textiles

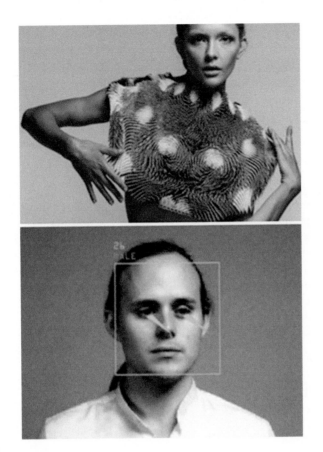

FIGURE 4.5

3D printed fashion. "Caress of the Gaze" by Behnaz Farahi multimaterial 3D printed clothing with smart textiles that adapt to the gaze of the observer.

Reproduced with permission from Farahi B. Material behaviours in 3D-printed fashion items.
Archit Des 2017;87:84–91.

and their drapability [65]. Danit Peleg has explored a variety of patterns and textures and noted that some can produce garments that hug the body or move in unique ways [59]. A variety of other designers have explored different patterns, as well as 3D printing of various structures on top of traditional fabrics, thus creating unique hybrids. In particular, auxetic patterns, which produce a negative poisson ratio, create unique possibilities [62]. Traditional materials narrow in one axis when stretched in the perpendicular direction: these patterns extend when stretched. All of these new possibilities will expand the options designers have for producing custom fashions.

While there are many possibilities for 3D printing of textiles, a number of companies have attempted to produce 3D printed textiles with varying results. While a number of companies have failed to produce viable products at a practical cost, Cosiflex by Tamicare is a 3D printed nonwoven textile that is currently used in 3D printed underwear [66]. While success has been limited to date, this will most likely improve as technology costs come down.

3D PRINTED FASHION

Due to the cost and expertise required, most actual 3D printed clothing to date has been limited to the world of high fashion. A variety of shows over the last 3—4 years have featured 3D printed dresses and other garments. Of note, Danit Pelig has created a variety of 3D printed fashions and has made an informative TED talk on the potential of 3D printed fashion [59]. Shows including threeAS-FOUR and Farahi Bezahi's biomimicry illustrate the potential of these garments [60,62]. However, even within these demonstrations, 3D printed materials often define the limitations of the application. While articles produced from flexible materials have produced interesting designs, many are still bulky and those produced from rigid materials require open cell frame or spring-like designs to provide the flexibility to accommodate the movement of the model [62]. However, as the materials and cost continue to improve, there is no doubt 3D printed clothing will create more possibilities.

REFERENCES

[1] Liaw C-Y, Guvendiren M. Current and emerging applications of 3D printing in medicine. Biofabrication 2017;9(2):024102. https://doi.org/10.1088/1758-5090/aa7279.

[2] Standard terminology for additive manufacturing—General principles—Terminology. ASTM International.

[3] Labrien D. Does 3D printing have a future for consumers. 2016. Available from: https://tech.co/3d-printing-future-consumers-2016-06.

[4] Langnau L. Testing consumer 3D printers. 2015. Available from: http://www.makepartsfast.com/testing-consumer-3d-printers/.

[5] Bull G, Haj-Hariri H, Atkins R, Moran P. An educational framework for digital manufacturing in schools. 3D Print Addit Manuf 2015;2(2):42—9. https://doi.org/10.1089/3dp.2015.0009.

[6] Moore F. 3D printing: separating science fiction from reality. Available from: http://airshipdaily.com/blog/3d-printing-separating-science-fiction-from-reality.

[7] Smith R. 8 hot 3D printing trends to watch in 2016. 2016. Available from: https://www.forbes.com/sites/ricksmith/2016/01/12/8-hot-3d-printing-trends-to-watch-in-2016/#7aca157a1d42.

[8] Pallottino F, Hakola L, Costa C, Antonucci F, Figorilli S, Seisto A, Menesatti P. Printing on food or food printing: a review. Food Bioprocess Technol 2016;9(5):725—33. https://doi.org/10.1007/s11947-016-1692-3.

[9]   Derossi A, Caporizzi R, Azzollini D, Severini C. Application of 3D printing for customized food. A case on the development of a fruit-based snack for children. J Food Eng 2018;220:65−75. https://doi.org/10.1016/j.jfoodeng.2017.05.015.

[10] Lipton JI, Cutler M, Nigl F, Cohen D, Lipson H. Additive manufacturing for the food industry. Trends Food Sci Technol 2015;43(1):114−23. https://doi.org/10.1016/j.tifs.2015.02.004.

[11] Godoi FC, Prakash S, Bhandari BR. 3d printing technologies applied for food design: status and prospects. J Food Eng 2016;179:44−54. https://doi.org/10.1016/j.jfoodeng.2016.01.025.

[12] Regueiro J, Negreira N, Simal-Gándara J. Challenges in relating concentrations of aromas and tastes with flavor features of foods. Crit Rev Food Sci Nutr 2015;57(10):2112−27. https://doi.org/10.1080/10408398.2015.1048775.

[13] Sun J, Zhou W, Huang D, Fuh JYH, Hong GS. An overview of 3D printing technologies for food fabrication. Food Bioprocess Technol 2015;8(8):1605−15. https://doi.org/10.1007/s11947-015-1528-6.

[14] Sun J, Zhou W, Yan L, Huang D, Lin L-y. Extrusion-based food printing for digitalized food design and nutrition control. J Food Eng 2018;220:1−11. https://doi.org/10.1016/j.jfoodeng.2017.02.028.

[15] Kouzani AZ, Adams S, Oliver R, Nguwi YY, Hemsley B, Balandin S. 3D printing of a pavlova. In: 2016 IEEE region 10 conference (TENCON); 2016/11. IEEE; 2016.

[16] Kouzani AZ, Adams S, Whyte JD, Oliver R, Hemsley B, Palmer S, Balandin S. 3D printing of food for people with swallowing difficulties. KnE Eng 2017;2(2):23. https://doi.org/10.18502/keg.v2i2.591.

[17] Hamilton CA, Alici G, in het Panhuis M. 3D printing vegemite and marmite: redefining "breadboards". J Food Eng 2018;220:83−8. https://doi.org/10.1016/j.jfoodeng.2017.01.008.

[18] Liu Z, Zhang M, Bhandari B, Yang C. Impact of rheological properties of mashed potatoes on 3D printing. J Food Eng 2018;220:76−82. https://doi.org/10.1016/j.jfoodeng.2017.04.017.

[19] Lille M, Nurmela A, Nordlund E, Metsä-Kortelainen S, Sozer N. Applicability of protein and fiber-rich food materials in extrusion-based 3D printing. J Food Eng 2018;220:20−7. https://doi.org/10.1016/j.jfoodeng.2017.04.034.

[20] Khaled SA, Burley JC, Alexander MR, Yang J, Roberts CJ. 3D printing of tablets containing multiple drugs with defined release profiles. Int J Pharm 2015;494(2):643−50. https://doi.org/10.1016/j.ijpharm.2015.07.067.

[21] Hsiao W-K, Lorber B, Reitsamer H, Khinast J. 3D printing of oral drugs: a new reality or hype? Expert Opin Drug Deliv 2017;15(1):1−4. https://doi.org/10.1080/17425247.2017.1371698.

[22] Kyobula M, Adedeji A, Alexander MR, Saleh E, Wildman R, Ashcroft I, Gellert PR, Roberts CJ. 3D inkjet printing of tablets exploiting bespoke complex geometries for controlled and tuneable drug release. J Contr Release 2017;261:207−15. https://doi.org/10.1016/j.jconrel.2017.06.025.

[23] Food and Drug Administration (FDA). Encyclopedia of immunotoxicology. Berlin, Heidelberg: Springer; 2015. p. 333.

[24] Dawood A, Marti BM, Sauret-Jackson V, Darwood A. 3D printing in dentistry. Br Dent J 2015;219(11):521−9. https://doi.org/10.1038/sj.bdj.2015.914.

[25] Norman J, Madurawe RD, Moore CMV, Khan MA, Khairuzzaman A. A new chapter in pharmaceutical manufacturing: 3D-printed drug products. Adv Drug Deliv Rev 2017; 108:39−50. https://doi.org/10.1016/j.addr.2016.03.001.

[26] Adams JW. 3D printing in anatomical and palaeontological teaching and research. Faseb J 2016;30(1 Suppl.). https://doi.org/10.1096/fasebj.30.1_supplement.92.2. 92.2.

[27] Sandström CG. The non-disruptive emergence of an ecosystem for 3D Printing— insights from the hearing aid industry's transition 1989−2008. Technol Forecast Soc Change 2016;102:160−8. https://doi.org/10.1016/j.techfore.2015.09.006.

[28] George M, Aroom KR, Hawes HG, Gill BS, Love J. 3D printed surgical instruments: the design and fabrication process. World J Surg 2016;41(1):314−9. https://doi.org/10.1007/s00268-016-3814-5.

[29] Zhu W, Ma X, Gou M, Mei D, Zhang K, Chen S. 3D printing of functional biomaterials for tissue engineering. Curr Opin Biotechnol 2016;40:103−12. https://doi.org/10.1016/j.copbio.2016.03.014.

[30] O'Brien EK, Wayne DB, Barsness KA, McGaghie WC, Barsuk JH. Use of 3D printing for medical education models in transplantation medicine: a critical review. Curr Transpl Rep 2016;3(1):109−19. https://doi.org/10.1007/s40472-016-0088-7.

[31] Administration USFDA. Draft guidance for industry and 6 Food and Drug Administration Staff, May 10, 2016. 2016.

[32] Birtchnell T, Urry J. A new industrial future?: 3D printing and the reconfiguring of production, distribution, and consumption. Routledge; 2016. 146 p.

[33] Craig Giffi BG. 3D opportunity for the automotive industry. Deloitte Insights; 2014.

[34] Portolés L, Jordá O, Jordá L, Uriondo A, Esperon-Miguez M, Perinpanayagam S. A qualification procedure to manufacture and repair aerospace parts with electron beam melting. J Manuf Syst 2016;41:65−75. https://doi.org/10.1016/j.jmsy.2016.07.002.

[35] Schwartz M. How 3D printing is transforming the aerospace industry. 2017. Available from: https://blog.trimech.com/how-3d-printing-in-transforming-the-aerospace-industry.

[36] Tay YWD, Panda B, Paul SC, Mohamed NAN, Tan MJ, Leong KF. 3D printing trends in building and construction industry: a review. Virtual Phys Prototyp 2017;12(3):261−76. https://doi.org/10.1080/17452759.2017.1326724. PMID: WOS:000416052500008.

[37] Moon MA. San Francisco startup 3D printed a whole house in 24 hours. 2017. Available from: https://www.engadget.com/2017/03/07/apis-cor-3d-printed-house/.

[38] Chimene D, Lennox KK, Kaunas RR, Gaharwar AK. Advanced bioinks for 3D printing: a materials science perspective. Ann Biomed Eng 2016;44(6):2090−102. https://doi.org/10.1007/s10439-016-1638-y.

[39] Groll J, Boland T, Blunk T, Burdick JA, Cho D-W, Dalton PD, Derby B, Forgacs G, Li Q, Mironov VA, Moroni L, Nakamura M, Shu W, Takeuchi S, Vozzi G, Woodfield TBF, Xu T, Yoo JJ, Malda J. Biofabrication: reappraising the definition of an evolving field. Biofabrication 2016;8(1):013001. https://doi.org/10.1088/1758-5090/8/1/013001.

[40] Jose RR, Rodriguez MJ, Dixon TA, Omenetto F, Kaplan DL. Evolution of bioinks and additive manufacturing technologies for 3D bioprinting. ACS Biomater Sci Eng 2016; 2(10):1662−78. https://doi.org/10.1021/acsbiomaterials.6b00088.

[41] Vyas C, Pereira R, Huang B, Liu F, Wang W, Bartolo P. Engineering the vasculature with additive manufacturing. Curr Opin Biomed Eng 2017;2:1−13. https://doi.org/10.1016/j.cobme.2017.05.008.

[42] Highley CB, Rodell CB, Burdick JA. Direct 3D printing of shear-thinning hydrogels into self-healing hydrogels. Adv Mater 2015;27(34):5075−9. https://doi.org/10.1002/adma.201501234.

[43] Jia W, Gungor-Ozkerim PS, Zhang YS, Yue K, Zhu K, Liu W, Pi Q, Byambaa B, Dokmeci MR, Shin SR, Khademhosseini A. Direct 3D bioprinting of perfusable vascular constructs using a blend bioink. Biomaterials 2016;106:58−68. https://doi.org/10.1016/j.biomaterials.2016.07.038.

[44] Zhang YS, Arneri A, Bersini S, Shin S-R, Zhu K, Goli-Malekabadi Z, Aleman J, Colosi C, Busignani F, Dell'Erba V, Bishop C, Shupe T, Demarchi D, Moretti M, Rasponi M, Dokmeci MR, Atala A, Khademhosseini A. Bioprinting 3D microfibrous scaffolds for engineering endothelialized myocardium and heart-on-a-chip. Biomaterials 2016;110:45−59. https://doi.org/10.1016/j.biomaterials.2016.09.003.

[45] Bendtsen ST, Quinnell SP, Wei M. Development of a novel alginate-polyvinyl alcohol-hydroxyapatite hydrogel for 3D bioprinting bone tissue engineered scaffolds. J Biomed Mater Res A 2017;105(5):1457−68. https://doi.org/10.1002/jbm.a.36036.

[46] Liu H, Yang L, Zhang E, Zhang R, Cai D, Zhu S, Ran J, Bunpetch V, Cai Y, Heng BC, Hu Y, Dai X, Chen X, Ouyang H. Biomimetic tendon extracellular matrix composite gradient scaffold enhances ligament-to-bone junction reconstruction. Acta Biomater 2017;56:129−40. https://doi.org/10.1016/j.actbio.2017.05.027.

[47] Cunniffe GM, Gonzalez-Fernandez T, Daly A, Sathy BN, Jeon O, Alsberg E, Kelly DJ. Three-dimensional bioprinting of polycaprolactone reinforced gene activated bioinks for bone tissue engineering. Tissue Eng A 2017. https://doi.org/10.1089/ten.tea.2016.0498.

[48] Keriquel V, Oliveira H, Rémy M, Ziane S, Delmond S, Rousseau B, Rey S, Catros S, Amédée J, Guillemot F, Fricain J-C. In situ printing of mesenchymal stromal cells, by laser-assisted bioprinting, for in vivo bone regeneration applications. Sci Rep 2017;7(1). https://doi.org/10.1038/s41598-017-01914-x.

[49] Murphy C, Kolan K, Li W, Semon J, Day D, Leu M. 3D bioprinting of stem cells and polymer/bioactive glass composite scaffolds for tissue engineering. Int J Bioprint 2017; 3(1). https://doi.org/10.18063/ijb.2017.01.005.

[50] Nguyen DG, Pentoney SL. Bioprinted three dimensional human tissues for toxicology and disease modeling. Drug Discov Today Technol 2017;23:37−44. https://doi.org/10.1016/j.ddtec.2017.03.001.

[51] Zhu W, Harris BT, Zhang LG. Gelatin methacrylamide hydrogel with graphene nanoplatelets for neural cell-laden 3D bioprinting. In: 2016 38th annual international conference of the IEEE Engineering in Medicine and Biology Society (EMBC); 2016/08. IEEE; 2016.

[52] Adams SD, Ashok A, Kanwar RK, Kanwar JR, Kouzani AZ. Integrated 3D printed scaffolds and electrical stimulation for enhancing primary human cardiomyocyte cultures. Bioprinting 2017;6:18−24. https://doi.org/10.1016/j.bprint.2017.04.003.

[53] Duan B. State-of-the-art review of 3D bioprinting for cardiovascular tissue engineering. Ann Biomed Eng 2016;45(1):195−209. https://doi.org/10.1007/s10439-016-1607-5.

[54] Bhise NS, Manoharan V, Massa S, Tamayol A, Ghaderi M, Miscuglio M, Lang Q, Shrike Zhang Y, Shin SR, Calzone G, Annabi N, Shupe TD, Bishop CE, Atala A, Dokmeci MR, Khademhosseini A. A liver-on-a-chip platform with bioprinted hepatic spheroids. Biofabrication 2016;8(1):014101. https://doi.org/10.1088/1758-5090/8/1/014101.

[55] Zhu W, Holmes B, Glazer RI, Zhang LG. 3D printed nanocomposite matrix for the study of breast cancer bone metastasis. Nanomed Nanotechnol Biol Med 2016;12(1): 69—79. https://doi.org/10.1016/j.nano.2015.09.010.

[56] Arrigoni C, Gilardi M, Bersini S, Candrian C, Moretti M. Bioprinting and organ-on-chip applications towards personalized medicine for bone diseases. Stem Cell Rev Rep 2017;13(3):407—17. https://doi.org/10.1007/s12015-017-9741-5.

[57] Cachon GP, Swinney R. The value of fast fashion: quick response, enhanced design, and strategic consumer behavior. Manag Sci 2011;57(4):778—95. https://doi.org/10.1287/mnsc.1100.1303.

[58] Ha-Brookshire JE, Hawley JM. Envisioning the clothing and textile-related discipline for the 21st century its scientific nature and domain from the global supply chain perspective. Cloth Text Res J 2013;31(1):17—31. https://doi.org/10.1177/0887302x12470024. PMID: WOS:000313643200002.

[59] Peleg D. Forget shopping. Soon you'll download you new clothes. TED Talks; 2015.

[60] Jacobson R.. The shattering truth of 3d-printed clothing. In: Wired.com, editor.

[61] Sun D. 3D printing in fashion promises to be huge—so what's holding us back?. 2016. Available from: https://www.usnews.com/news/stem-solutions/articles/2016-12-09/3-d-printing-in-fashion-promises-to-be-huge-so-whats-holding-us-back.

[62] Farahi B. Material behaviours in 3d-printed fashion items. Archit Des 2017;87(6): 84—91. https://doi.org/10.1002/ad.2242. PMID: WOS:000414347700012.

[63] Heater B. Adidas join Carbon's board as its 3D printed shoes finally drop. 2018. Available from: https://techcrunch.com/2018/01/18/adidas-joins-carbons-board-as-its-3d-printed-shoes-finally-drop/.

[64] Perry A. 3D-printed apparel and 3D-printer: exploring advantages, concerns, and purchases. Int J Fash Des Technol Edu 2018;11(1):95—103. https://doi.org/10.1080/17543266.2017.1306118.

[65] Kinematics dress. 2014.

[66] Perez D. 3D-printed underwear is now a reality. 2013. Available from: http://www.ubergizmo.com/2013/11/3d-printed-underwear-is-now-a-reality/.

Complex Congenital Heart Disease

5

Kevin Luke Tsai[1], Subhi J. Al'Aref[2,3], Alexander R. van Rosendael[3], Jeroen J. Bax[4]

The Brooklyn Hospital Center, Brooklyn, New York, NY, United States[1]; Department of Radiology, Weill Cornell Medicine, New York, NY, United States[2]; Dalio Institute of Cardiovascular Imaging, NewYork-Presbyterian Hospital, New York, NY, United States[3]; Leiden University Medical Center, Leiden, The Netherlands[4]

INTRODUCTION

Three-dimensional (3D) printing technology is now a valuable tool to complement the traditional methods used for visualizing cardiovascular anatomy, to assist clinical decision-making, and to plan and simulate surgical and percutaneous procedures. Nowhere is this more valuable than in congenital heart disease (CHD), which frequently manifests in unique anatomies that are difficult to study with more conventional imaging techniques. An increasing number of patients with CHD survive into adulthood owing to advances in neonatal screening and surgical treatment; consequently, the adult population with CHD is constantly growing [1] and the need for precise and personalized interventions to treat these patients have made them one of the most studied for the applications of 3D printed models [2]. They have also been used for preoperative planning of surgeries and percutaneous interventions, potentially reducing costs, time, and complications, and are valued educational tools that help patients and their families better understand the anatomical features of particular cardiovascular diseases. 3D printing has the potential to revolutionize the delivery of individualized treatments, especially in the broad and complex group of congenital cardiovascular pathologies, and in the future, it will play an essential role in the manufacturing of implantable and anatomically accurate prostheses, positioners, or molds comprised of the patient's own cells.

CARDIOVASCULAR VISUALIZATION

The creation of 3D printed cardiac models provides the opportunity to gain tactile and visual familiarity of anatomical structures. Creation of such models relies on high-quality imaging, most commonly with cardiac magnetic resonance imaging (CMR), but cardiac computed tomography (CCT) is frequently used when patients have devices that preclude a CMR or if more image resolution is required for the 3D model [3]. Each imaging modality has different strengths—CT enhances visualization of extracardiac anatomy [4], CMR is superior for quantification of ventricular volumes and myocardial architecture [5], and 3D echocardiography is best for visualization of valve morphology [6]. The selection of the optimal imaging modality for use in creating 3D printed models is extremely important and the utilization of numerous imaging methods to fabricate one model has been used to create those of high accuracy [7]. In CHD, the manifestations of disease that fall under one broad category often include many unique anatomic variations; so visualizing their anatomical features and relationships to nearby structures can prove challenging with conventional imaging techniques. 3D imaging using CT, CMR, and echocardiography are commonly used, but 3D printed models provide even more information, being tangible objects that can be manipulated—cut, bent, and rotated—to optimize viewing planes, and allow visualization of complete 3D anatomic features of specific pathologies, enabling a greater understanding of complex anatomies. The visual and tactile features of a 3D printed model are advantageous for simulating surgeries and percutaneous interventions, permitting more precise planning of highly complex procedures, and in turn providing the ability to better appreciate potential difficulties and to predict the likelihood of success or failure [8]. These models rapidly and intuitively convey complex anatomic arrangements that are patient-specific and are valuable in aiding medical professionals, students, and patients to better understand unique structural heart conditions.

Biglino et al. [9] fabricated a 3D printed model of the heart and main vessels of an 11-year-old boy with a repaired truncus arteriosus, right aortic arch, pulmonary artery stenosis, and conduit stenosis from CT data. This model facilitated communication among physicians, the patient, and family members. The patient and his family found the 3D model to be more informative in terms of explaining the underlying anatomical derangements compared to traditional medical imaging, while the surgeons and cardiologists noted that they gained a better understanding of the complex relationship between the narrowed right pulmonary artery and the aorta, which ultimately guided the preoperative decision-making process and surgical strategy (Fig. 5.1).

Jaworski et al. [10] reported a 6-month-old boy with pulmonary atresia and a ventricular septal defect (VSD) with complete agenesis of the pulmonary trunk and absence of the right and left pulmonary arteries. As a newborn, he had shunts from the ascending aorta to remnants of the underdeveloped pulmonary arteries. The team initially performed CT to determine a way to proceed with reconstruction of the right ventricular (RV) outflow tract, but because of the complexity of the case and the need to more clearly delineate anatomical structures for presurgical planning, a 3D printed model was created. This model was useful in allowing the multidisciplinary team to plan the appropriate treatment approach and to examine potential pitfalls and difficulties, while also proving beneficial in providing the parents better insight into their child's disease and facilitated discussions regarding possible treatment strategies.

Medical education and training is an important use of 3D printed cardiovascular models. Because of the unique manifestations of CHD, a model can expose medical

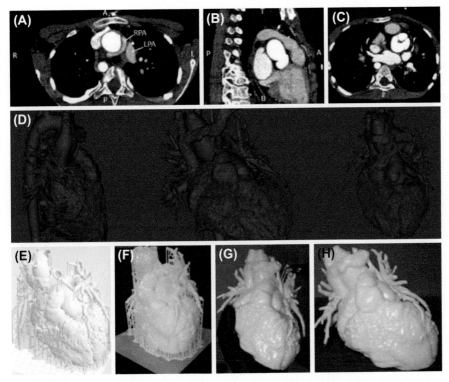

FIGURE 5.1

A summary of the steps in 3D printing. Computed Tomography images are shown in axial and sagittal planes (A and B respectively), with the position of the right and left pulmonary arteries in relation to the aorta highlighted. Structures of interest are subsequently segmented (*red outline* in panel (C)) and the images are then rendered in 3D (panel D). A scaffold (E) is built around the model prior to 3D printing (F) and then is removed (G), resulting in the final patient-specific model (H).

Reproduced with permission from Biglino G, et al. The perception of a three-dimensional-printed heart model from the perspective of different stakeholders: a complex case of truncus arteriosus. Front Pediatr 2017;5:209.

students to the wide variability that exists within a single condition. These models are useful both in the clinical setting and in the classroom, and their visual and tactile nature allows students to physically manipulate these models at will, helping them understand and retain complex anatomical defects. The utility of these models in medical education translates well to medical training. Currently, acquiring the necessary skills for percutaneous or surgical procedures is extremely costly in terms of finances and time, in addition to potentially putting patients at risk. Because the prevalence of complex CHDs is rare, procedural training in managing them is mostly opportunity-based, and these cases are difficult to encounter during medical training (or even during an individual's career). 3D printed cardiovascular models provide avenues to acquire and perfect the necessary specialized skills by presenting opportunities to practice and rehearse rare and challenging cases (and in some instances, even develop new techniques). Accurate models of highly complex diseases can be created using this technology, which has permitted simulation-based training to expand over the last few decades and allowed proceduralists to efficiently refine their skills in managing patients with complex diseases and comorbidities in standardized environments [11]. One study by Shirakawa et al. [12] involved performing detailed morphological assessments and preoperative simulations using a 3D-printed cardiac model for surgical planning in a rare case of double-chambered right ventricle (DCRV) and found that the similarity between the printed heart and actual heart was excellent (Fig. 5.2).

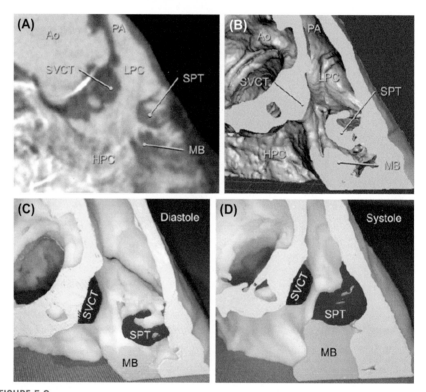

FIGURE 5.2

Illustration of 3D printing in a patient with DCRV. Cross-sectional images showing right ventricular outflow tract stenosis and the surrounding structures on cardiac computed tomography (A) and 3D polygonal data (B). Patient-specific 3D DCRV models are printed in diastole (C) and systole (D). *Ao,* aorta; *DCRV,* double-chambered right ventricle; *HPC,* high-pressure inflow chamber; *LPC,* low-pressure outflow chamber; *MB,* moderator band; *PA,* pulmonary artery; *SPT,* septoparietal trabecula; *SVCT,* supraventricular crest.

Reproduced with permission from Shirakawa T, et al. Morphological analysis and preoperative simulation of a double-chambered right ventricle using 3-dimensional printing technology. Interact Cardiovasc Thorac Surg 2016;22:688–90.

PROCEDURAL PLANNING

CHDs that are routinely managed in the cardiac catheterization laboratory include atrial septal defects (ASDs), patent ductus arteriosus, and muscular VSDs. Although simpler cardiac defects can be understood by traditional two-dimensional (2D) imaging modalities, many patients with CHD have more complex anatomies and often require 3D imaging and modeling to ensure adequate preprocedural preparation. Traditional forms of 3D imaging display models on 2D screens, but full understanding of complex anatomies and spatial relationships might be hampered due to the inability to obtain tactile information and manipulate cardiac models at will [13].

3D printing has quickly become indispensable in CHD for both percutaneous and surgical planning and training. Some of the most frequently printed pathologies include ASDs and VSDs, which are used to assist with intraoperative spatial navigation and patch sizing [8]. Thus, 3D printed cardiovascular models assist proceduralists in narrowing the types and sizes of devices needed for patient-specific pathologies and facilitate preprocedural simulation, thereby reducing the overall cost of percutaneous or surgical procedures [3].

Patients scheduled to undergo percutaneous or surgical procedures with utilization of 3D printed models as part of the preprocedural planning likely experience reduced radiation and contrast exposure. Because a 3D printed model was able to clarify certain parts of their patient's complex anatomy, Ryan et al. [14] were able to help their patient avoid preoperative cardiac catheterization. Another benefit of 3D printed cardiac models is that they can be created to be radiopaque. In the case of a patient with hypoplastic transverse aortic arch, Valverde et al. [15] fabricated a 3D cardiac model from CMR data of a repaired aortic coarctation in a 15-year-old boy with hypoplastic aortic arch. Under fluoroscopic guidance, simulation of the endovascular stenting of the hypoplastic arch was carried out, and the procedure was subsequently performed successfully on the patient. The authors found that there was excellent agreement between the 3D printed model and both CMR and X-ray angiographic images. The valuable role that 3D printed models play in clinical decision-making was further illustrated by Valverde et al. [16], who evaluated the impact of 3D printed models in a prospective study involving 10 international centers and 40 patients with complex CHD. The study concluded that there was agreement among 96% of the surgeons, who surveyed that 3D models provided both better understanding of CHD morphology and improved surgical planning, and in almost half of the complex cases, 3D printed models helped redefine the surgical approach.

EXAMPLES OF 3D PRINTING IN CONGENITAL HEART DISEASE
SEPTAL DEFECTS

3D printing has been used to aid in the navigation of occluder devices in patients with ASDs. ASDs and VSDs are among the most common CHDs and because they can occur in combination with other cardiac pathologies, their anatomies are

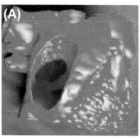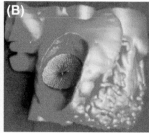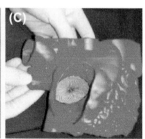

FIGURE 5.3

Utility of 3D printing in challenging atrial septal defect situations. A personalized 3D printed heart model was produced from computed tomography images (A), and in vitro occlusion was achieved successfully with a 38-mm Amplatzer septal occluder (B and C).

Reproduced with permission from Chaowu Y, et al. Three-dimensional printing as an aid in transcatheter closure of secundum atrial septal defect with rim deficiency: in vitro trial occlusion based on a personalized heart model. Circulation 2016;133:e608–10.

frequently complex. Traditionally, hemodynamically relevant defects have been surgically closed, but increasingly, percutaneous closure has become a safe alternative [17], and Olivieri et al. [18] have provided evidence that 3D printing of VSD anatomy correlates very well with conventional 2D echocardiographic measurements.

Chaowu et al. [19] described a case of a 54-year-old woman who was found to have a large (30 mm) ASD with left-to-right shunt with poor septal rim. Because large ASDs are sometimes associated with deficient rims, device implantation can be difficult and increases the risk of device embolization. Although surgical repair was recommended, the patient elected to undergo transcatheter closure. An elastic rubber heart model was 3D printed using multislice CT data and in vitro trial occlusion was performed preoperatively, which assisted the interventionalists in precisely sizing the occluder device and decreasing the risk of complications (Fig. 5.3). The authors also noted that these models could be used to prevent unnecessary transcatheter closure and potentially screen appropriate candidates for transcatheter ASD closure.

Kim et al. [20] described a case in which a 30-year-old man was referred for percutaneous closure of a congenital muscular VSD. CCT angiography revealed a large inferoseptal VSD (14×12 mm). This source image was used to generate a 3D physical model, which allowed the physicians to appropriately size the occluder device. Using the printed model, the interventionalists simulated placement of various catheters to establish the optimal approach to crossing the VSD and were able to determine that a Judkins right 4.0 cm diagnostic catheter from the retrograde aortic approach was the safest method. The patient subsequently underwent a percutaneous VSD closure with the use of a 12-mm AMPLATZER occluder device without complications (Fig. 5.4).

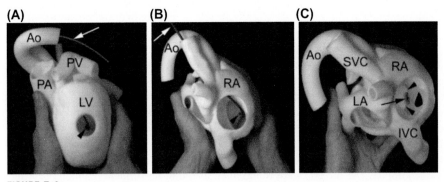

FIGURE 5.4

Use of a 3D printed model in a patient with a muscular VSD. Panel A shows crossing the ventricular septum (*arrowhead*) with a JR-4 diagnostic catheter. Panel B shows the catheter (*arrow*) pathway to the VSD (*arrowhead*) via the SVC approach. Panel C is a view from the RA into the RV illustrating prominent septal curvature (*arrowheads*) and the position of Amplatzer occluder across the interventricular septum (*arrow*). *Ao*, aorta; *IVC*, inferior vena cava; *JR*, Judkins right; *LV*, left ventricle; *PV*, pulmonary veins; *RA*, right atrium; *RV*, right ventricle; *SVC*, superior vena cava; *VSD*, ventricular septal defect.

Reproduced with permission from Kim MS, et al. Rapid prototyping: a new tool in understanding and treating structural heart disease. Circulation 2008;117:2388–94.

DOUBLE OUTLET RIGHT VENTRICLE

In double outlet right ventricle (DORV), more than half of both semilunar valves of the great arteries arise from the morphologic right ventricle, and this condition is almost always accompanied by a VSD [21]. These patients have wide variability in the 3D relationships among the ventricles, ventricular septum, and great arteries and include diverse clinical manifestations secondary to the changes in infundibular and intracardiac morphologies [8]. The unique spatial arrangement of these structures governs the type and approach of intervention that will be undertaken. Although traditional 2D imaging (e.g., echocardiography) can represent these structures fairly well, Farooqi et al. [22] showed that 3D printed models can offer valuable anatomic information that is difficult to obtain with traditional imaging modalities. The team described a 7-year-old boy with DORV and noncommitted VSD who had undergone a bidirectional Glenn anastomosis at 6 months of age. At that time, he was considered a poor candidate for biventricular repair and was palliated toward a single-ventricle pathway. A CMR was performed due to the complexity of his case, but the ability to determine the 3D spatial arrangement of the intracardiac structures was limited when displayed on a 2D screen. Based on the 3D printed model that was then produced and the valuable spatial information that was gathered from it, the cardiologists and surgeons deemed the patient to be a good candidate for biventricular repair and were able to visualize a realistic potential baffle pathway. The patient underwent a successful operation to transition from single-ventricle palliation to two-ventricle repair. Predischarge echocardiography revealed no baffle obstruction and normal biventricular function. The 3D printed cardiac model played a vital role in determining the patient's candidacy for a biventricular repair while helping to eliminate ambiguities in defining the patient's cardiovascular anatomy.

Garekar et al. [23] performed CCT and CMR scans on five patients with DORV and generated 3D printed models. These models were compared with the surgeons' direct observations intraoperatively and with echocardiography, CCT, and CMR images. The accuracy of the models was scored based on a unique scale created for the purpose of the study, and involved scoring by a radiologist, a surgeon, and a cardiologist. The models were found to be useful in presurgical planning, scoring higher than conventional imaging with respect to surface spatial orientation and intracardiac anatomy. Facilitating multidisciplinary involvement and communication was also noted to be an advantage of developing the 3D printed models.

Bhatla et al. [24] utilized 3D printed cardiac models to better understand the anatomy of a patient with DORV. Patient-specific anatomy was printed as multiple, flexible, whole-heart models using a PolyJet printer (a rapid prototyping process that uses additive manufacturing) and a supplemental layer of myocardium was added outside the blood pool, which allowed the surgeon to perform a realistic surgical simulation to visualize the 3D relationships among structures from various planes and to determine the optimal surgical approach. Previous 2D imaging had revealed that the newborn had a doubly committed VSD intersected by a large conal septum, which made potential surgical VSD closure challenging. The 3D printed model enabled presurgical simulation, better visualization of structural relationships, and as a result, improved preprocedural preparation (Fig. 5.5).

TRANSPOSITION OF THE GREAT ARTERIES

Dextro-transposition of the great arteries (d-TGA) is a potentially lethal form of CHD in newborns and infants and consists of the aorta originating from the morphologic right ventricle and the pulmonary artery originating from the morphologic left ventricle, resulting in the pulmonary and systemic circulations being connected in parallel rather than in series.

As previously described, 3D printed cardiac models have been used in surgical simulations and are extremely helpful for diseases with complex anatomies that are difficult to fully grasp with traditional 2D or 3D imaging modalities, and corrective surgeries in CHD pose special challenges due to the heterogeneity of malformation and broad spectrum of anatomical spatial interrelationships [25]. Valverde et al. [25] constructed a model of the heart of a 1.5-year-old boy with TGA, VSD, and pulmonary stenosis using CMR data to assist with surgical planning. This model was used in the evaluation of the location and dimensions of the VSD, as well as its relationship with the aorta and pulmonary artery, and helped the surgeons confirm that the patient would benefit from the Nikaidoh, rather than the Rastelli, procedure (Fig. 5.6). This process allowed the team to determine the

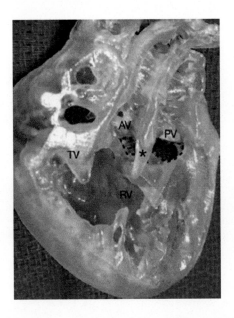

FIGURE 5.5

3D printing in DORV. A coronal cross-section through a 3D printed model (with the RV free wall removed) showing the relationship between the conal septum (*black asterisk*), aortic and pulmonic valve. Highlighted too is the plane of a ventricular septal defect (outlined by *black hashed line*). *AV*, aortic valve; *DORV*, double outlet right ventricle; *PV*, pulmonic valve; *RV*, right ventricle; *TV*, tricuspid valve.

Reproduced with permission from Bhatla P, et al. Surgical planning for a complex double-outlet right ventricle using 3D printing. Echocardiography 2017;34:802–4.

extent of the subaortic septum that needed to be excised in order to translocate the aorta posteriorly and to establish the location for placing the intraventricular patch to close the VSD. Because the surgeons were able to utilize the 3D printed model in presurgical planning, no unexpected findings were encountered during the operation. The spatial relationships among the VSD, origin of the aorta, and stenotic pulmonary artery were found to be identical to those produced by the model. Olivieri et al. [26], during a case of pulmonary venous baffle obstruction in a patient who had undergone post-Mustard operation for d-TGA, attempted multiple catheter approaches to the pulmonary venous atrium on a 3D printed cardiovascular model, taking advantage of its radiopaque property. After the optimal approach was selected and stent deployment was simulated in the model, the patient underwent successful stent deployment in the stenosed pulmonary venous baffle (Fig. 5.7). These cases clearly illustrate the critical role that 3D printed models play in preoperative assessment—the printed models facilitate precise decision-making and anatomical orientation, helping the teams avoid improvisation and potentially decreasing intra-operative time and complications.

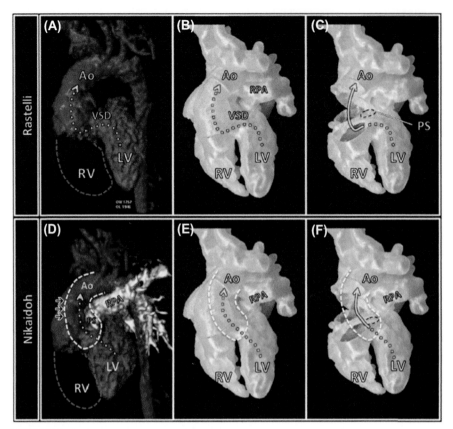

FIGURE 5.6

3D Printing in transposition of the great arteries, comparing the Rastelli and Nikaidoh procedures. The Rastelli procedure may result in suboptimal hemodynamics as seen on reconstructed MRI images (panel A), which was subsequently confirmed on 3D printed models revealing a likelihood of left ventricular outflow tract obstruction (panels B and C). The Nikaidoh procedure, on the other hand, may achieve a more laminar flow to the aorta as suggested by reconstructed MRI images (panel D). Such a finding was confirmed by a 3D printed model (panels E and F) which permitted eventual planning of the resection of the intraventricular outlet septum. *Ao,* Aorta; *LPA,* left pulmonary artery; *LV,* left ventricle; *MRI,* magnetic resonance imaging; *PS,* pulmonary stenosis; *RA,* right atrium; *RPA,* right pulmonary artery; *RV,* right ventricle; *VSD,* ventricular septal defect.

Reproduced with permission from Valverde I, et al. Three-dimensional patient-specific cardiac model for surgical planning in Nikaidoh procedure. Cardiol Young 2015;25:698–704.

3D printed cardiac models can also aid in the perioperative planning of adults with CHD, those who were treated from a younger age [27]. In a complicated case of a patient diagnosed with TGA, pulmonary atresia, large VSD, ASD, tricuspid regurgitation, and dextrocardia (and who had already undergone prior surgical

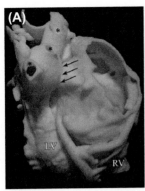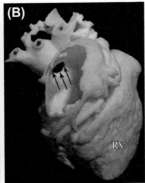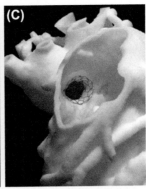

FIGURE 5.7

3D Printing in transposition of the great arteries, guiding percutaneous therapy of an obstructed pulmonary venous baffle. (A) 3D printed model from the right side showing three right pulmonary veins (*asterisks*) entering the pulmonary venous atrium posteriorly. The stenosis (*arrows*) is located between the dilated posterior pulmonary venous atrium (to left) and the much larger anterior atrial chamber (to right). (B) Enface view, after removal of the anterior atrial wall, of the stenotic orifice (*arrows*). (C) View of the stent in the stenotic orifice.

Reproduced with permission from Olivieri L, et al. 3D heart model guides complex stent angioplasty of pulmonary venous baffle obstruction in a Mustard repair of D-TGA. Int J Cardiol 2014;172:e297–8.

interventions), 3D CMR data was used to create virtual and physical 3D cardiac models to more precisely locate the VSD and to obtain views of the heart that would have been impossible intraoperatively. The virtual models were viewed with red-blue glasses for 3D orientation, and virtual cuts were made—cuts that could not have been performed intraoperatively—to better orient the team to intracardiac structures. The physical models were taken into the operating room and used for orientation of the coronary arteries and abnormal structures at the opened situs. This process allowed the surgeons to perform several procedures that would not have been risked if only traditional imaging data were available (without a physical model). The authors concluded that the physical and virtual models were clearly advantageous in preoperative preparation and improved the team's orientation of the heart and its structures intraoperatively.

TETRALOGY OF FALLOT

The four components of tetralogy of Fallot (ToF) are an outlet VSD, obstruction to RV outflow, overriding of the aorta (>50%), and RV hypertrophy. Although the dominant site of RV outflow obstruction is at the subvalvular level, the outflow tract can be atretic in some instances. Deferm et al. [28] discussed a case involving an adult female born with complex ToF, consisting of pulmonary atresia and major aortopulmonary collateral arteries (MAPCAs). This patient had undergone unifocalization of the left- and right-sided MAPCAs, which were connected by a left- and right-sided modified Blalock-Taussig shunt to the systemic circulation (Fig. 5.8). During a later repair, a pulmonary homograft was implanted on the RV outflow tract and connected to the left- and right-sided unifocalized MAPCAs using the previously applied two small Blalock-Taussig conduits. Because the two conduits were now undersized, she required redo surgery to combat the pressure load on the right ventricle (80 mm Hg, estimated on Doppler echocardiography). The initial chest CT images were challenging for understanding the pulmonary circulation and its relationship with the bronchial tree, so a 3D printed cardiovascular model was fabricated using the CT data. This model was used in preprocedural planning and in discussions with the patient regarding the treatment plan. She underwent surgical repair, which consisted of replacing the pulmonary homograft and conduits, and inserting two extra conduits from a new homograft to the distal pulmonary vascular tree. A second 3D printed model was created after the surgery, which clearly showed that all conduits were well connected, and repeat echocardiography revealed that her RV systolic pressure had significantly decreased to 50 mm Hg. The two 3D printed cardiac models were noted to have enhanced patient-physician communication when used to explain the underlying pathology and planned surgical intervention, and later on when discussing surgical outcome with the team and the patient.

Ryan et al. [14] used a 3D printed model using CCT to plan placement of a central aortopulmonary shunt and subsequent coiling of redundant collateral vessels

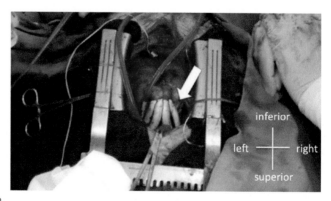

FIGURE 5.8

An example of a female patient born with complex tetralogy of Fallot consisting of pulmonary atresia and MAPCAs requiring redo surgery to correct increasing RV pressure overload due to undersized conduits. Perioperative images show two new left sided conduits and two new right sided conduits originating from the replaced homograft (*white arrow*). *MAPCAs*, major aortopulmonary collateral arteries; *RV*, right ventricle.

Reproduced with permission from Deferm S, et al. 3D-printing in congenital cardiology: from flatland to spaceland. J Clin Imaging Sci 2016;6:8.

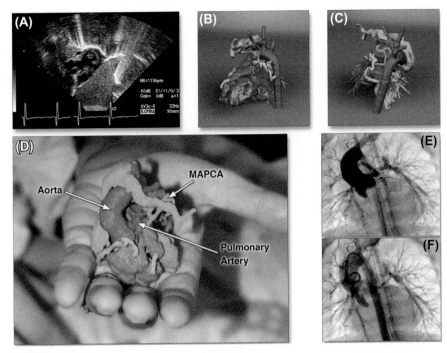

FIGURE 5.9

3D printing in ToF. (A) Transthoracic echocardiography shows the presence of ToF/MAPCAs/pulmonary atresia. (B and C) 3D reconstruction illustrates spatial relationship of patient-specific geometry (pulmonary arteries (*blue*), aorta (*red*), and MAPCAs (*green and yellow*)) for central aortopulmonary shunt placement and coil planning. (E and F) Invasive coronary angiography captured prior to placement of MAPCA embolization coils. *MAPCAs*, major aortopulmonary collateral arteries; *ToF*, tetralogy of Fallot.

Reproduced with permission from Ryan JR, et al. A novel approach to neonatal management of tetralogy of Fallot, with pulmonary atresia, and multiple aortopulmonary collaterals. JACC Cardiovasc Imaging 2015;8:103–4.

in a patient with ToF and pulmonary atresia. This model was made available intraoperatively and was useful in conveying pulmonary vascular anatomy and identifying MAPCAs (Fig. 5.9). Use of this model spared the patient from having to undergo preoperative cardiac catheterization, freeing him from general anesthesia, radiation, and contrast exposure.

EVALUATION FOR VENTRICULAR ASSIST DEVICES

Approximately one-quarter of adults with CHD will progress to heart failure by 30 years of age [29]. Unfortunately, the use of ventricular assist devices (VAD) has remained rare due to the variable anatomy and complex physiology in this patient population. Farooqi et al. [30] explored the use of 3D printed cardiac models in patients with TGA to determine whether such models could be helpful in identifying hurdles and selecting optimal locations for VAD placement. Barriers to VAD placement in these patients are numerous, including the RV apex not being as well developed as the LV apex, RV dilatation causing severe structural distortion, the presence of trabeculations and the moderator band (which can cause inflow obstruction), scarring of the mediastinum, and abnormal hemodynamics. Furthermore, in patients whose anatomy is not amenable to biventricular circulation, the presence of Fontan circuit leaks or stenosis must be assessed, as well as any stenosis of the pulmonary arteries.

Farooqi et al. [30] provided several clinical examples that were illustrative. A 36-year-old man with d-TGA who had undergone a Mustard procedure at 5 years of age developed severe RV failure. A 3D printed model from a CMR dataset revealed prominent trabeculations of the systemic RV and a moderator band, both of which could have caused inflow obstruction. Similarly, a 51-year-old man with l-TGA demonstrated clinical signs of heart failure for 1 year and his systemic right ventricle was found to be severely hypertrophied and moderately dilated, with systolic and diastolic dysfunction. A 3D printed model was created using CT data, which revealed prominent systemic RV trabeculations and an anterior and leftward aorta (Fig. 5.10). The third example was that of a 37-year-old woman with tricuspid atresia and d-TGA who had undergone Fontan palliation at 7 years of age. She required a Maze procedure for atrial arrhythmias and underwent a Fontan revision at 28 years of age. She continued to have significant ascites and atrial arrhythmias, so a 3D printed model from CT images was created, which clearly revealed the location of the Fontan pathway and its spatial relationship with the rest of the structures. These cases highlight the ability of 3D printed models to accurately portray complex anatomies and elucidate both unfavorable and potentially favorable areas for VAD placement, which would lower the barriers to offering VAD therapy by potentially identifying more candidates and extending this therapy to those who would derive the most benefit.

EVALUATION FOR HEART TRANSPLANTATION

3D printed models can enhance and facilitate perioperative planning in cardiac transplantation. Many CHDs are extremely challenging to visualize using traditional methods and many variations of broadly categorized diseases are rarely encountered. Because the number of patients with CHD who survive into adulthood are increasing, the need for cardiac transplantation is also increasing [31,32]. Heart transplantation is an extremely complex procedure in CHD patients due to their complex anatomy and resulting pathophysiological maladaptations. These patients also experience longer waiting times and have higher mortality than other transplantation candidates [33].

Smith et al. [34] demonstrated the role that 3D printed models could provide in the perioperative planning of heart transplantation in a patient with complex CHD and situs inversus dextrocardia. This scenario was faced with special technical

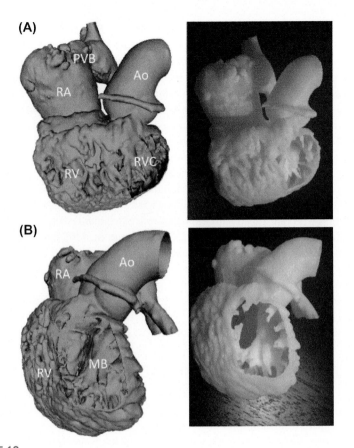

FIGURE 5.10

3D printing in heart failure and congenital heart disease. 3D rendered images and the corresponding printed model in a 51-year-old patient with I-transposition of the great arteries and heart failure, seen in the anterior (A) and leftward (B) aspect. Prominent trabeculations in the systemic right ventricle and an anterior and leftwards aorta are illustrated, which assisted with planned surgical planning. *Ao*, aorta; *MB*, moderator band; *PVB*, pulmonary venous baffle; *RA*, right atrium; *RV*, right ventricle; *RVC*, right ventricular cavity.

Reproduced with permission from Farooqi KM, et al. 3D printing to guide ventricular assist device placement in adults with congenital heart disease and heart failure. JACC Heart Fail 2016;4:301—11.

challenges, as the vascular structures of a donor heart—in the orientation of a normal heart—had to be anastomosed into a cavity that has been adapted for a heart with situs inversus. The surgical team decided to create a color-coded 3D printed model of the patient's heart using contrast-enhanced, multiphase ECG-gated CT data to optimize comprehension of the spatial relationships and orientations among the involved structures in advance of the surgery. The authors noted several main benefits of the 3D model that were specific to that case. First, the model helped the team appreciate the necessity of harvesting a longer donor ascending aorta than what is customary due to the relationship between the transposed ascending aorta and main pulmonary artery visualized on the model. Second, the model allowed advanced recognition of the anomalous drainage of the four pulmonary veins to a separate confluence behind the native heart, which drained superiorly to a left-sided superior vena cava (SVC). This avoided significant intraoperative dissection and exploration, allowing the surgeons to ligate the connection of the anomalous venous confluence to the left SVC prior to anastomosing the confluence posterior to the heart directly to the donor left atrium. Finally, the model enabled advanced recognition of the shortened left SVC, which meant that both the SVC and brachiocephalic veins were harvested with the donor heart.

For patients who fail staged palliation after univentricular repair, orthotopic heart transplantation is the definitive therapy [35]. Sodian et al. [36] fabricated replicas from CT and MRI angiograms of the hearts of two patients with failed palliation of hypoplastic left heart syndrome. Like most patients with CHD, their anatomies were extremely complex. These 3D printed models not only assisted the surgeons in visualizing their patients' unique anatomies to precisely plan procedures and anticipate complications, but were also valuable in allowing the explant team to determine the dimensions and requirements of the donor heart.

USE IN RESOURCE-LIMITED ENVIRONMENTS

Kappanayil et al. [37] showed that this technology is both feasible and particularly valuable in resource-limited environments. In India (where the authors noted that access and expertise for complex heart surgery are limited), life-sized 3D cardiac models were printed for five patients with complex CHD. Noteworthy was that these patients had previously undergone extensive clinical evaluations and cardiovascular imaging in the past, but no plans for definitive procedures were ever planned given the complexity of their underlying anatomical substrates. Two patients had complex DORV, two patients had crisscross AV connections, and one patient had a congenitally corrected TGA with pulmonary atresia. All patients underwent CMR, while one patient underwent additional CCT imaging. The authors concluded that these 3D printed prototypes enabled improved understanding of anatomy, identification of technical challenges, and much more precise surgical planning. All five of these patients eventually underwent successful corrective operations.

CURRENT LIMITATIONS

There are currently no long-term trials evaluating the utility, safety, overall costs, and outcomes of 3D printing technology in clinical practice, but published experiences of handling 3D printed models, utilization for surgical simulations, and testing implants show that this technology has a promising future. Data showing that 3D printed models enhance preoperative planning in CHD have largely been based on case reports and studies using self-reported data that might not directly translate to patient care. As a result, more research is needed to determine this technology's true clinical impact. Some studies have been encouraging—Schievano et al. [38] showed that 3D printed models were able to correctly reproduce anatomical details within a few millimeters, with excellent correlation between the 3D images and the printed models. However, multicenter trials to evaluate the accuracy of these models should still be undertaken so that this technology can become increasingly validated while universal standards and guidelines can be developed to ensure consistent quality. Operator skill, choice of imaging modality, and choice of segmentation algorithm must be closely examined to avoid bias. Confounding variables including image quality and differences among imaging techniques; operator variability in image acquisition must also be accounted for. These trials would need to include a wide range of source imaging modalities, 3D printing methods, and cardiovascular modeling scenarios, as 3D printed models are increasingly used for clinical decision-making and modeling errors could have profound consequences. As 3D printing technology becomes more utilized, costs will decrease while processing times are improved, which will ensure further research and development and lead to more widespread availability. Continued development of more useful, complex, and comprehensive cardiac models via 3D printing should continue while the overall costs and benefits to patients are regularly assessed.

3D printed cardiovascular models are static and cannot reproduce physiologic changes that occur during the cardiac cycle, which is limiting when used to assess structural changes during variable hemodynamic conditions. Although 3D printed cardiac models allow visual and tactile representations of complex anatomy, currently virtual 3D models must be utilized to visualize the heart at various stages of the cardiac cycle. An ideal 3D printed model would be able to mimic the anatomic and physiologic changes that occur during the cardiac cycle, allowing the incorporation of hemodynamic data to unique pathologies to predict the results of percutaneous and surgical procedures. This would avoid the creation of multiple models, during distinct stages of the cardiac cycle, for one lesion—a potentially time- and resource-consuming process [39]. Efforts underway hold promise. Maragiannis et al. [40] printed a flexible LVOT in a patient with severe aortic stenosis and then replicated physiologic hemodynamics by attaching the printed aorta to a pulsatile flow imaging circuit. Catheter measurement and Doppler assessment confirmed that the circuit was able to duplicate the extent of the obstruction. Additionally, functional flow models have recently been used for functional assessment under different in vitro flow conditions and are extremely useful to model valvular pathology [41].

THE FUTURE OF MEDICAL 3D PRINTING

3D printing holds the promise of revolutionizing the processes in which CHD is approached and understood. The models derived from this technology have already been used in clinical decision-making and will herald a new era of personalized medicine. In the present, efforts should be made at utilizing a multimodal approach to produce the most accurate representation of a patient's specific anatomy, as these models have already shown their crucial roles in directing clinical decision-making [42]. Each specific imaging modality has its own strengths, as discussed previously, and choices made by the medical team for use in 3D printed models have potential to affect procedural approaches and patient outcomes.

Material development is an area that has seen remarkable advances. The first 3D printed models were rigid structures that were useful for visualization of anatomic relationships of specific structures. Since then, more flexible materials have been developed to better replicate the mechanics of cardiac tissue, such as the Tango and Vero families of materials (Stratasys Ltd.) that can create models with a broad range of stiffness and compliance specifications [40,43]. PolyJet machines can replicate very complex anatomic structures and can create models that combine multiple colors and materials simultaneously [44] and are especially useful when visualizing the mechanics of structures, such as valves, or to replicate anatomical elements with different tissue characteristics. This is also extremely important when fabricating functional models for experiments in pressurized flow loops with tailored hemodynamic conditions [40]. Material blends are needed to model pathological conditions, such as calcific structures within cardiac valves, and to model a narrow range of cardiovascular structures and behavior during different hemodynamic conditions. This approach was demonstrated by Gosnell et al. [45] in a patient with congenitally corrected TGA by integrating CT and 3D transesophageal echocardiography for hybrid 3D printing using HeartPrint Flex technology (Materialise N.V.), a material that mimics human cardiovascular tissue. Further data regarding the behavior of these structures, both normal and pathological, would be required, in addition to further research into more 3D print material blends. Mechanical testing of a range of materials with comparisons to tissues of mitral valve leaflets has shown that they are only capable of mimicking a small range of possible mitral valve leaflet deformations [41], preventing fully accurate modeling of human cardiac tissue. The field of material research is continuing to rapidly progress and will inevitably result in further development of printed cardiovascular tissue that behaves almost identically to human cardiac tissue and fabrication of implantable, 3D printed, patient-specific cardiac prostheses utilizing novel materials and blends.

One group of techniques known as bioprinting involves the process of laying down cells in a controlled manner using 3D printing techniques, with the goal of having those cells maintain their viability in their new environments. Bracaglia et al. [46] developed a 3D printed degradable hybrid scaffold by combining polyethylene glycol and homogenized pericardium matrix to achieve a biological environment that encouraged recognition by native cells. The authors hypothesized that this hybrid hydrogel could reduce inflammatory signals from macrophages and nearby endothelial cells, thereby impacting the healing environment at the implantation site—in this case, a vascular graft. The authors then developed a cytocompatible printing method using this hybrid hydrogel as a scaffold—a promising result that could lead to future graft development. This active area of research has displayed success in tissue engineering of biocompatible materials that support tissue formation and growth of cells and other supporting components, with recent advances including 3D printing of blood vessels and heart valves [47−49]. Further research should eventually lead to 3D bioprinting of implantable, patient-specific, cardiac or vascular structures, allowing this technology to have a direct role in disease management and realizing the dream of truly personalized cardiac medicine.

To make complex CHDs more accessible for medical professionals and students, 3D libraries have been created to enable sharing of files that can be freely accessed and printed on 3D printers by individuals and other institutions—another important step in improving medical education and increasing the availability of this technology. Several online 3D libraries have already been established, including NIH 3D Print Exchange (https://3dprint.nih.gov/) and 3D Hope Medical (http://www.3dhopemedical.com/), with the hopes of preserving 3D anatomic data and democratizing access to 3D CHD pathology. As the costs of 3D printing decrease and the availability of open-source medical imaging software increases, these libraries will have a significant impact on the practice of CHD by facilitating collaboration and increasing proficiency, in addition, to becoming an essential resource for learning, practicing, and procedural development.

An emerging technology that shows promise for being complementary to physical 3D printed cardiovascular models is 3D holographic imaging, which can be utilized intraprocedurally. These models, being virtual, are beneficial because they allow dynamic rendering of various cutting planes to optimize viewing angles. Beitnes et al. [50] demonstrated that it was feasible to analyze 3D echocardiographic data on a holographic 3D display during mitral valve interventions and that the holographic display achieved good representation of all mitral valve segments, offering additional perspectives on this condition. Since then, the team has continued work on developing a process for real-time streaming of echocardiographic data to the 3D holographic display, potentially enabling its use during invasive procedures.

CONCLUSION

Medical imaging has significantly advanced over the past several decades, and the emergence of 3D printing technology heralds a revolution in modern medical imaging. Applications of this technology include medical education, visualization of highly complex anatomy in CHDs, facilitation of patient-physician communication, and preoperative planning by providing a substrate for anatomic orientation and procedural simulations. 3D printing technology has undergone rapid advances and will eventually lead to the development of patient-specific implantable devices that can be populated with patients' own cells. There are still barriers to overcome before its adoption becomes more widespread, such as identifying proper indications that optimize its benefits and development of standards to increase efficiency and decrease costs. Progress in material science, more accurate functional flow models, and 3D bioprinting has shown that we are moving in the right direction. Online 3D libraries are available that will encourage broad availability and use. It is not difficult to imagine that the technology of 3D cardiovascular printing will soon be part of routine clinical practice, reshaping and challenging our current views on personalized, precise, and affordable cardiovascular medicine.

REFERENCES

[1] Benziger CP, Stout K, Zaragoza-Macias E, Bertozzi-Villa A, Flaxman AD. Projected growth of the adult congenital heart disease population in the United States to 2050: an integrative systems modeling approach. Popul Health Metr 2015;13:29.

[2] Giannopoulos AA, Mitsouras D, Yoo SJ, Liu PP, Chatzizisis YS, Rybicki FJ. Applications of 3D printing in cardiovascular diseases. Nat Rev Cardiol 2016;13:701—18.

[3] Grant EK, Olivieri LJ. The role of 3-d heart models in planning and executing interventional procedures. Can J Cardiol 2017;33:1074—81.

[4] Goitein O, Salem Y, Jacobson J, et al. The role of cardiac computed tomography in infants with congenital heart disease. Isr Med Assoc J 2014;16:147—52.

[5] Luijnenburg SE, Robbers-Visser D, Moelker A, et al. Intra-observer and interobserver variability of biventricular function, volumes and mass in patients with congenital heart disease measured by CMR imaging. Int J Cardiovasc Imaging 2010;26:57—64.

[6] Black D, Vettukattil J. Advanced echocardiographic imaging of the congenitally malformed heart. Curr Cardiol Rev 2013;9:241—52.

[7] Noecker AM, Chen JF, Zhou Q, et al. Development of patient-specific three-dimensional pediatric cardiac models. ASAIO J 2006;52:349—53.

[8] Giannopoulos AA, Steigner ML, George E, Barile M, Hunsaker AR, Rybicki FJ, Mitsouras D. Cardiothoracic applications of 3-dimensional printing. J Thorac Imaging 2016;31:253—72.

[9] Biglino G, Moharem-Elgamal S, Lee M, Tulloh R, Caputo M. The perception of a three-dimensional-printed heart model from the perspective of different Stakeholders: a complex case of truncus arteriosus. Front Pediatr 2017;5:209.

[10] Jaworski R, Haponiuk I, Chojnicki M, Olszewski H, Lulewicz P. Three-dimensional printing technology supports surgery planning in patients with complex congenital heart defects. Kardiol Pol 2017;75:185.

[11] Park CS, Rochlen LR, Yaghmour E, Higgins N, Bauchat JR, Wojciechowski KG, Sullivan JT, McCarthy RJ. Acquisition of critical intraoperative event management skills in novice anesthesiology residents by using high-fidelity simulation-based training. Anesthesiology 2010;112:202—11.

[12] Shirakawa T, Koyama Y, Mizoguchi H, Yoshitatsu M. Morphological analysis and preoperative simulation of a double-chambered right ventricle using 3-dimensional printing technology. Interact Cardiovasc Thorac Surg 2016;22:688—90.

[13] Hadeed K, Acar P, Dulac Y, Cuttone F, Alacoque X, Karsenty C. Cardiac 3D printing for better understanding of congenital heart disease. Archives of cardiovascular diseases. Arch Cardiovasc Dis 2018;111:1—4.

[14] Ryan JR, Moe TG, Richardson R, Frakes DH, Nigro JJ, Pophal S. A novel approach to neonatal management of tetralogy of Fallot, with pulmonary atresia, and multiple aortopulmonary collaterals. JACC Cardiovasc Imaging 2015;8:103—4.

[15] Valverde I, Gomez G, Coserria JF, Suarez-Mejias C, Uribe S, Sotelo J, Velasco MN, Santos De Soto J, Hosseinpour AR, Gomez-Cia T. 3D printed models for planning endovascular stenting in transverse aortic arch hypoplasia. Catheter Cardiovasc Interv 2015;85:1006—12.

[16] Valverde I, Gomez-Ciriza G, Hussain T, Suarez-Mejias C, Velasco-Forte MN, Byrne N, Ordoñez A, Gonzalez-Calle A, Anderson D, Hazekamp MG, Roest AA. Three-dimensional printed models for surgical planning of complex congenital heart defects: an international multicentre study. Eur J Cardio Thorac Surg 2017;52:1139—48.

[17] Baumgartner H, Bonhoeffer P, De Groot NM, de Haan F, Deanfield JE, Galie N, Gatzoulis MA, Gohlke-Baerwolf C, Kaemmerer H, Kilner P, Meijboom F, Mulder BJ, Oechslin E, Oliver JM, Serraf A, Szatmari A, Thaulow E, Vouhe PR, Walma E. ESC Guidelines for the management of grown-up congenital heart disease. Eur Heart J 2010;31:2915—57.

[18] Olivieri LJ, Krieger A, Loke YH, Nath DS, Kim PC, Sable CA. Three-dimensional printing of intracardiac defects from three-dimensional echocardiographic images: feasibility and relative accuracy. J Am Soc Echocardiogr 2015;28:392—7.

[19] Chaowu Y, Hua L, Xin S. Three-dimensional printing as an aid in transcatheter closure of secundum atrial septal defect with rim deficiency: in vitro trial occlusion based on a personalized heart model. Circulation 2016;133:e608—10.

[20] Kim MS, Hansgen AR, Wink O, Quaife RA, Carroll JD. Rapid prototyping: a new tool in understanding and treating structural heart disease. Circulation 2008;117:2388—94.

[21] Wilcox BR, et al. Surgical anatomy of double-outlet right ventricle with situs solitus and atrioventricular concordance. J Thorac Cardiovasc Surg 1981;82:405—17.

[22] Farooqi KM, Gonzalez-Lengua C, Shenoy R, Sanz J, Nguyen K. Use of a three dimensional printed cardiac model to assess suitability for biventricular repair. World J Pediatr Congenit Heart Surg 2016;7:414—6.

[23] Garekar S, Bharati A, Chokhandre M, Mali S, Trivedi B, Changela VP, Solanki N, Gaikwad S, Agarwal V. Clinical application and multidisciplinary assessment of three dimensional printing in double outlet right ventricle with remote ventricular septal defect. World J Pediatr Congenit Heart Surg 2016;7:344—50.

[24] Bhatla P, Tretter JT, Chikkabyrappa S, Chakravarti S, Mosca RS. Surgical planning for a complex double-outlet right ventricle using 3D printing. Echocardiography 2017;34:802—4.

[25] Valverde I, Gomez G, Gonzalez A, Suarez-Mejias C, Adsuar A, Coserria J, Uribe S, Gomez-Cia T, Hosseinpour A. Three-dimensional patient-specific cardiac model for surgical planning in Nikaidoh procedure. Cardiol Young 2015;25:698—704.

[26] Olivieri L, Krieger A, Chen MY, Kim P, Kanter JP. 3D heart model guides complex stent angioplasty of pulmonary venous baffle obstruction in a Mustard repair of D-TGA. Int J Cardiol 2014;172:e297−8.

[27] Mottl-Link S, Hübler M, Kühne T, Rietdorf U, Krueger JJ, Schnackenburg B, De Simone R, Berger F, Juraszek A, Meinzer HP, Karck M, Hetzer R, Wolf I. Physical models aiding in complex congenital heart surgery. Ann Thorac Surg 2008;86:273−7.

[28] Deferm S, Meyns B, Vlasselaers D, Budts W. 3D-printing in congenital cardiology: from flatland to spaceland. J Clin Imaging Sci 2016;6:8.

[29] Norozi K, Wessel A, Alpers V, Arnhold JO, Geyer S, Zoege M, Buchhorn R. Incidence and risk distribution of heart failure in adolescents and adults with congenital heart disease after cardiac surgery. Am J Cardiol 2006;97:1238−43.

[30] Farooqi KM, Saeed O, Zaidi A, Sanz J, Nielsen JC, Hsu DT, Jorde UP. 3D printing to guide ventricular assist device placement in adults with congenital heart disease and heart failure. JACC Heart Fail 2016;4:301−11.

[31] Karamlou T, Hirsch J, Welke K, Ohye RG, Bove EL, Devaney EJ, Gajarski RJ. A United Network for Organ Sharing analysis of heart transplantation in adults with congenital heart disease: outcomes and factors associated with mortality and retransplantation. Thorac Cardiovasc Surg 2010;140:161−8.

[32] Ross HJ, Law Y, Book WM, Broberg CS, Burchill L, Cecchin F, Chen JM, Delgado D, Dimopoulos K, Everitt MD, Gatzoulis M, Harris L, Hsu DT, Kuvin JT, Martin CM, Murphy AM, Singh G, Spray TL, Stout KK. Transplantation and mechanical circulatory support in congenital heart disease: a scientific statement from the American Heart Association. Circulation 2016;133:802−20.

[33] Stewart GC, Mayer Jr JE. Heart transplantation in adults with congenital heart disease. Heart Fail Clin 2014;10:207−18.

[34] Smith ML, McGuinness J, O'Reilly MK, Nolke L, Murray JG, Jones JF. The role of 3D printing in preoperative planning for heart transplantation in complex congenital heart disease. Ir J Med Sci 2017;186:753−6.

[35] d'Udekem Y, Iyengar AJ, Cochrane AD, Grigg LE, Ramsay JM, Wheaton GR, Penny DJ, Brizard CP. The Fontan procedure: contemporary techniques have improved long-term outcomes. Circulation 2007;116:I1157−64.

[36] Sodian R, Weber S, Markert M, Loeff M, Lueth T, Weis FC, Daebritz S, Malec E, Schmitz C, Reichart B. Pediatric cardiac transplantation: three-dimensional printing of anatomic models for surgical planning of heart transplantation in patients with uni-ventricular heart. J Thorac Cardiovasc Surg 2008;136:1098−9.

[37] Kappanayil M, Koneti NR, Kannan RR, Kottayil BP, Kumar K. Three-dimensional-printed cardiac prototypes aid surgical decision-making and preoperative planning in selected cases of complex congenital heart diseases: early experience and proof of concept in a resource-limited environment. Ann Pediatr Cardiol 2017;10:117.

[38] Schievano S, Migliavacca F, Coats L, Khambadkone S, Carminati M, Wilson N, Deanfield JE, Bonhoeffer P, Taylor AM. Percutaneous pulmonary valve implantation based on rapid prototyping of right ventricular outflow tract and pulmonary trunk from MR data. Radiology 2007;242:490−7.

[39] Meier LM, Meineri M, Qua Hiansen J, Horlick EM. Structural and congenital heart disease interventions: the role of three-dimensional printing. Neth Heart J 2017;25:65−75.

[40] Maragiannis D, Jackson MS, Igo SR, Schutt RC, Connell P, Grande-Allen J, Barker CM, Chang SM, Reardon MJ, Zoghbi WA, Little SH. Replicating patient-specific severe aortic valve stenosis with functional 3D modeling. Circ Cardiovasc Imaging 2015;8:e003626.

[41] Vukicevic M, Puperi DS, Jane Grande-Allen K, Little SH. 3D printed modeling of the mitral valve for catheter-based structural interventions. Ann Biomed Eng 2016;44: 3432.

[42] Kurup HK, Samuel BP, Vettukattil JJ. Hybrid 3D printing: a game-changer in personalized cardiac medicine? Expert Rev Cardiovasc Ther 2015;13:1281−4.

[43] Little SH, Vukicevic M, Avenatti E, Ramchandani M, Barker CM. 3D printed modeling for patient-specific mitral valve intervention: repair with a clip and a plug. JACC Cardiovasc Interv 2016;9:973−5.

[44] Vukicevic M, Mosadegh B, Min JK, Little SH. Cardiac 3D printing and its future directions. JACC Cardiovasc Imaging 2017;10:171−84.

[45] Gosnell JM, Pietila T, Samuel BP, Vettukattil JJ. Hybrid three-dimensional printing derived from multiple imaging modalities. In: Catheter interventions in congenital, structural, and valvular heart disease; 2015 June 24−27; Frankfurt (Germany); 2015.

[46] Bracaglia LG, Messina M, Winston S, Kuo CY, Lerman M, Fisher JP. 3D printed pericardium hydrogels to promote wound healing in vascular applications. Biomacromolecules 2017;18:3802−11.

[47] Hockaday LA, Kang KH, Colangelo NW, Cheung PY, Duan B, Malone E, Wu J, Girardi LN, Bonassar LJ, Lipson H, Chu CC, Butcher JT. Rapid 3D printing of anatomically accurate and mechanically heterogeneous aortic valve hydrogel scaffolds. Biofabrication 2012;4:035005.

[48] Kucukgul C, Ozler SB, Inci I, Karakas E, Irmak S, Gozuacik D, Taralp A, Koc B. 3D bioprinting of biomimetic aortic vascular constructs with self-supporting cells. Biotechnol Bioeng 2015;112:811−21.

[49] Lueders C, Jastram B, Hetzer R, Schwandt H. Rapid manufacturing techniques for the tissue engineering of human heart valves. Eur J Cardio Thorac Surg 2014;46: 593−601.

[50] Beitnes JO, Klboe LG, Karlsen JS, Urheim S. Mitral valve analysis using a novel 3D holographic display: a feasibility study of 3D ultrasound data converted to a holographic screen. Int J Cardiovasc Imaging 2015;31:323−8.

Valvular Heart Disease

6

Qusai Saleh, John Moscona, Thierry Le Jemtel

Tulane University School of Medicine, New Orleans, Louisiana, United States

INTRODUCTION

Three-dimensional (3D) printing refers to anatomic modeling of high-quality volumetric images obtained by computed tomography (CT) angiography, cardiac magnetic resonance imaging (CMR), or transesophageal/thoracic echocardiography (TEE, TTE). Cardiovascular 3D printing is particularly suited for the study and management of cardiac valve pathology. The application of 3D printing to the personalized treatment of patients with cardiac valve pathology is rapidly progressing. However, while 3D printing undoubtedly provides unique information it has not been yet demonstrated to affect clinical outcome.

With 3D printing technology, anatomic modeling can be fabricated from different types of textured materials. The variable pliability and strength of printing materials can mimic native tissue. Initially, anatomic modeling was fabricated from rigid polymers. The 3D printing technology has progressed with the use of deformable materials and 3D bioprinting where cells, biomolecules, and biomaterials are dispensed layer by layer [1]. Materials used in 3D bioprinting remain to be tested against native cardiovascular tissue.

Anatomic modeling with 3D printing, which can elucidate complex cardiac pathology, helps assess cardiac valve function, plan structural interventions, and test percutaneous structural devices [2–8]. Specifically, anatomic modeling with 3D printing has been applied to periprocedural planning of transcatheter interventions for mostly the aortic and mitral valves [2–4]. Anatomic modeling with 3D printing complements cadaver-based learning of cardiac anatomy, particularly in the presence of congenital heart disease [5–7].

Whether cardiac valve repair/replacement is surgical or transcatheter, anatomic modeling with 3D printing can simulate the procedure and therefore help with patient selection, procedural technique, and avoidance of complications [8]. The present chapter reviews the technical aspects of anatomic modeling with 3D printing. The applications of 3D printing technology to the current management of aortic, mitral, tricuspid, and pulmonic cardiac valve diseases are then considered.

3D Printing Applications in Cardiovascular Medicine. https://doi.org/10.1016/B978-0-12-803917-5.00006-7

IMAGING CONSIDERATIONS

Image acquisition is the first step of 3D printing. Anatomic modeling with 3D printing requires volumetric imaging data that can be acquired by 3D echocardiography, CT, and CMR [9]. Cardiac valves are best imaged by echocardiography due to rapid motion of the valve leaflets [10]. Echocardiography is widely available, relatively inexpensive, and devoid of radiation exposure [8,11]. In some instances, 3D TEE provides all the necessary data to create patient-specific accurate models of cardiac valve leaflets. For the mitral valve (MV), deep transgastric TEE imaging of the papillary muscles can be digitally combined with data from the midesophageal view to model the entire MV apparatus [12]. Poor acoustic windows and artifact occasionally mitigate the use of echocardiography.

The complexity of cardiac valve disease and the proximity of key myocardial structures may require the use of multiple imaging modalities. Due to its excellent spatial resolution and clear delineation of different tissue types, CT imaging is routinely used for anatomic modeling with 3D printing. Like echocardiography, CMR is devoid of radiation exposure. However, its spatial resolution is less than that of CT. Overall, either CT or CMR are best suited for 3D printing of vascular lumen, as well as cardiac chambers [1]. Besides echocardiography, CT or CMR are quite useful in the planning of complex valvular interventions, as they provide highly detailed images of cardiac structures. Such a "hybrid" imaging strategy to create a 3D model has been utilized in the evaluation of congenital heart disease, and can be similarly applied to complex valvular pathology [13]. When severe calcifications hinder the visualization of a diseased valve with 3D TEE, cardiac CT can provide accurate volumetric 3D data for modeling the calcified valve pathology [14].

After imaging acquisition, the next step in 3D printing is image segmentation (Fig. 6.1). Segmentation is the process that converts datasets into a 3D digital model based on the separation of specific anatomic regions of interest. A major issue of the segmentation process is that manipulation of source data may distort true anatomy and pathology [15]. Particularly with small detailed structures such as cardiac valves, experience with segmentation software is essential for ensuring the accuracy of 3D anatomic modeling [16]. In the framework of MV reconstruction, image segmentation can be achieved from 3D TEE imaging data or from the combination of TEE and CT imaging data [12,17−19].

After image segmentation, 3D digital modeling is refined using computer-aided design (CAD) software, and is saved as a stereolithographic file for 3D printing (Fig. 6.1). Among the different types of available 3D printing technologies, stereolithography and PolyJet technology are the most commonly used in cardiac valve modeling. PolyJet technology offers extremely high-resolution 3D printing and is capable of using multiple materials and colors simultaneously [9]. It can produce highly complex models with smooth surfaces and thin walls (down to 0.016 mm) [20].

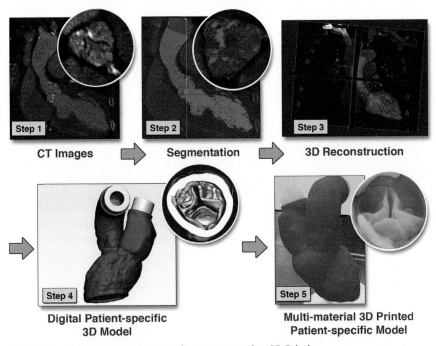

FIGURE 6.1 Modeling of patient-specific anatomy using 3D Printing.

Reproduced with permission from Vukicevic M, et al. Cardiac 3D printing and its future directions. JACC
Cardiovasc Imaging 2017;10:171–84.

APPLICATIONS IN VALVULAR HEART DISEASE

The most common use of 3D printing in anatomic modeling in cardiac valve disease
is periprocedural planning of complex aortic and MV interventions [20]. Anatomic
modeling with 3D printing can help with: (1) selection and sizing of the device; (2)
delineation of the device landing zones; and (3) prediction of device/valve interaction [21]. Although 3D imaging alone may provide some valuable information, how
the device may interact with the native tissue can only be gleaned from an accurate
physical 3D model. The most important applications of 3D printing will be discussed for each cardiac valve, with emphasis on transcatheter-based therapies for
aortic stenosis (AS) and mitral regurgitation.

AORTIC VALVE DISEASE

The prevalence of aortic valve disease increases with age. One population-based
study showed that the overall age-adjusted prevalence of moderate or severe
aortic valve disease is approximately 2.5% [22]. After mitral regurgitation, aortic

regurgitation (AR) is the second most prevalent cardiac valve disorder at 0.5%, followed by AS at 0.4%. However, the overall prevalence of AS in adults older than 75 years is about 2.8% [22].

Anatomical, genetic, and clinical factors all contribute to the pathogenesis of AS [23]. Valvular calcification leading to AS frequently occurs in patients with trileaflet aortic valves. Congenital bicuspid valves accounts for 60% of the patients with AS [24]. Rheumatic aortic valve disease resulting in AS is relatively uncommon in developed countries, but remains a major problem worldwide.

Although mortality is not increased in truly asymptomatic AS, in the absence of surgical or endovascular valve replacement, mortality is $\geq 50\%$ at 2 years, and $>80\%$ at 3 years in symptomatic patients [25,26]. Aortic valve replacement should be considered for symptomatic AS regardless of the patient's age at presentation, when overall life expectancy is >1 year, and the likelihood of survival is $>25\%$ at 2 years after valve replacement [23,27]. Currently, medical therapies cannot prevent or slow the progression of AS. Optimal management of AS depends on early diagnosis, accurate determination of stenosis severity, and appropriate timing and type of aortic valve replacement.

Transcatheter aortic valve replacement (TAVR) is increasingly performed in patients with severe AS. It is currently recommended as an alternative approach for patients not only with severe or prohibitive surgical risk, but also with intermediate risk associated with surgical aortic valve replacement [28]. Newer TAVR prostheses and the use of TAVR in lower risk cohorts are presently being investigated [29,30]. Whether based on safety or patient preference, TAVR provides patients with a reasonable alternative to surgical aortic valve replacement and is poised to become the predominant strategy to treat AS in the upcoming years [29,31].

As noted in patients with AS, bicuspid aortic valve and calcific valve disease are the most common causes of chronic AR in the United States and other developed countries [27]. In many developed countries, rheumatic heart disease remains the leading cause of AR [22]. Most often AR progresses slowly with steady volume overload promoting left ventricular (LV) dilation and hypertrophy [27]. Optimal management of AR highly depends on accurate determination of severity, and timing for valve replacement.

Surgical aortic valve replacement is at present the only approved mode of valve replacement in patients with native AR. In the absence of calcification, the annular segment of the ascending aorta dilates in patients with native AR. Annular dilatation is a relative contraindication for transcatheter valve replacement as it increases the risk of prosthesis embolization and paravalvular leak. The reported favorable outcomes of patients with severe AS after TAVR [28,29,31] have led to consider TAVR for the management of patients with severe AR (from native aortic valve regurgitation or failing regurgitant surgical valves) who are at high or prohibitive risk for surgery. However, experienced centers have recently reported worse results with TAVR for severe native AR than in patients with severe AS [32].

ANATOMY OF THE AORTIC VALVE

Appreciation of the aortic root relation to coronary arteries, MV, and conduction system is essential for safe implantation of aortic valve prostheses. Anatomic modeling with 3D printing is likely to help predict potential complications after insertion of a prosthetic valve.

Aortic Root and the Aortic "Annulus"

The aortic root is a direct continuation of the LV outflow tract (LVOT), extending from the basal attachment of the aortic valve leaflets within the LV to their distal attachment at the level of the sinutubular junction (STJ) [33]. The aortic root lies to the right and posterior relative to the sub-pulmonary infundibulum, with its posterior margin wedged between the orifice of the MV and the muscular ventricular septum [34]. Two-thirds of the lower part of the aortic root circumference is connected to the muscular ventricular septum, and one-third is in fibrous continuity with the anterior horn of the MV annulus.

An essential feature of the aortic root is the semilunar attachment of the valve leaflets, supported in crown-like fashion within its cylindrical shape. The basal attachment of the leaflets corresponds to the hemodynamic demarcation (a virtual ring) from ventricular to arterial pressures. The hinge lines of the leaflets cross the anatomic ventricular-arterial junction (a true ring), which is delineated by the transition from the LV myocardium to the fibroelastic valve sinuses [35] (Fig. 6.2). The top of the crown at the STJ (another true ring) is demarcated by the sinus ridge and the related sites of distal attachment of the leaflets [34].

The normal aortic root can be described as a truncated cone. When expressed in normal fixed human hearts as a percentage of the largest diameter at the level of the sinuses of Valsalva, the diameters of the inlet and outlet are 97% and 81%, respectively [42]. The aortic root is also dynamic, continuously changing shape throughout the cardiac cycle [34]. The change in diameter at the level of STJ and at the base of the valve is plus 12% in systole and minus 16% in diastole [36]. Such changes need to be considered when modeling or selecting valve prostheses [34].

The aortic root at the STJ is larger in patients with AS than in normal subjects [37]. However, the severity of AS does not affect aortic root diameters. Thus, aortic root dilation at the STJ may be an early adaptive response that does not progress [37]. Dilation of the ascending aorta is common in patients with AS, which needs to be addressed at the time of valve replacement [23]. The association between AS and aortic dilatation is further complicated by the phenotypic overlap between calcific AS and congenital bicuspid valve disease [23]. Patients with bicuspid aortic valves have larger aortic root diameters and different aortic root phenotypes than patients with trileaflet AS [38]. The most frequent pattern of dilation involves the tubular ascending aorta [39,40]. It progresses rapidly, irrespective of valve morphology and function [40].

Besides immediate postprocedural complications, oversizing a prosthetic valve can result in redundancy of leaflet tissue resulting in compressive and tensile stresses

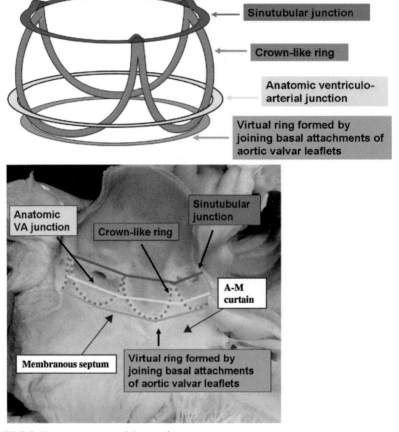

FIGURE 6.2 Normal anatomy of the aortic root.

For the purpose of transcatheter valve selection, the size of the aortic valve "annulus" at the level of basal attachments (*green line*) of the aortic valve leaflets dictates the size of the prosthesis.

Reproduced with permission from Piazza N, et al. Anatomy of the aortic valvar complex and its implications for transcatheter implantation of the aortic valve. Circ Cardiovasc Interv 2008;1:74–81.

that may alter function or durability of the prosthetic valve [41]. On another hand, undersizing the prosthesis may not fully alleviate AS. Accurate evaluation of the aortic root and selection of the appropriate sized prosthesis are essential steps in the management of patients with AS [42]. In regards to transcatheter valve selection, size of the aortic valve "annulus" at the level of basal attachments or hinge points of the aortic valve cusps dictates the size of the prosthesis [43] (Fig. 6.2). Accurate dimensions of the aortic annulus are important to curtail complications such as paravalvular regurgitation and migration of the implanted valve.

Aortic Valve Leaflets

As previously noted, aortic valve leaflets have semilunar attachments within the aortic root. Leaflets' geometry and size vary greatly from patient to patient and even in a given patient [34]. In bicuspid valves, the right and the left coronary cusps are fused in 70%–80% of patients and the right and noncoronary cusps are fused in 20%–30% of patients [38]. In the presence of severe calcifications, it may be difficult to differentiate a bicuspid from a trileaflet valve. Of note, bulky commissural calcifications are often present within the sinus of Valsalva. They may deform the aortic root leading to overestimation of the actual annular size in preprocedural planning for TAVR. In the presence of severe calcifications, multimodality imaging is most helpful [43].

Location of the Coronary Arteries

The ostia of the coronary arteries commonly arise within the two anterior sinuses of Valsalva and are often positioned just below the STJ [34]. It is not unusual for the coronary arteries to originate above the STJ [44]. Knowledge of the location of the coronary ostia is essential for appropriate percutaneous replacement of the aortic valve, as coronary arteries with low take-off are at risk of obstruction from the prosthesis itself or from the native aortic valve leaflets being crushed against the aortic wall [45].

Other features of aortic valve anatomy are important for sizing and choosing the appropriate device. The angle between aortic root and LVOT becomes obtuse with aging, whereas the root is in straight extension of the LVOT in young patients [34]. Asymmetrical septal hypertrophy may create an obstacle to proper seating of the aortic prosthesis. Lastly, the position of the atrioventricular node should also be kept in mind, as it sits in close proximity to the subaortic region and membranous septum of the LVOT. Proper seating and sizing can help decrease the risk of complete heart block or intraventricular conduction abnormalities [45].

3D PRINTING APPLICATIONS

Anatomic modeling with 3D printing is helpful at different stages of TAVR. In preprocedural planning, 3D printed models can identify appropriate positioning of aortic valve prostheses and lessen specific procedural risks such as an extremely calcified aortic root or possible aortic branch occlusion during valve deployment [2,3]. Selection of prosthesis size is of paramount importance in TAVR. However, the imaging gold standard for valve sizing has not yet been established. CT-measured aortic annulus diameters may better correlate with intraoperative measurements than TEE-derived measurements [46]. Overall, CT may overestimate and TEE may underestimate aortic valve annulus due to its elliptical structure [47]. In clinical practice, CT and TEE measurements are commonly averaged. Anatomic modeling with 3D printing may be used for simulated procedures in order to select appropriate prosthetic valve and optimal position [3]. Valve-in-valve procedures present some of the most challenging aortic cases encountered in clinical practice.

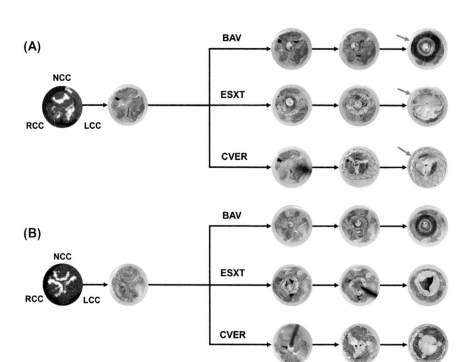

FIGURE 6.3 3D printed model simulations to investigate the effect of asymmetric calcium distribution of the aortic valve as a risk factor for the need of permanent pacemaker implantation after transcatheter aortic valve replacement.

Patient 1 (A) had heavy calcifications (*red*) in the LCC, which caused the balloon and THV devices to move away from the LCC towards RCC/NCC commissure adjacent to the conduction pathway. Patient 2 (B) did not have heavy calcifications of the LCC and therefore such lateral deflections of the devices were not seen. *BAV*, balloon valvuloplasty; *CVER*, Medtronic CoreValve Evolut R; *ESXT*, Edwards SAPIEN XT; *LCC*, left coronary cusp; *NCC*, non-coronary cusp; *RCC*, right coronary cusp; *THV*, transcatheter heart valve.

Reproduced with permission from Fujita B, et al. Calcium distribution patterns of the aortic valve as a risk factor for the need of permanent pacemaker implantation after transcatheter aortic valve implantation. Eur Heart J Cardiovasc Imaging 2016;17:1385–93.

Specific simulation algorithms that include 3D printing of an aortic annulus can evaluate optimal valve position and different valvular hemodynamic profiles to determine the most appropriate repair approach to valve-in-valve procedures [48]. Aortic root models by single-step 3D printing have also been used to test stented aortic valves and examine aortic root strain for preinterventional planning [49,50] (Fig. 6.3).

An important issue regarding 3D printed aortic valve models is how accurately they reproduce the anatomic and functional properties of severe AS. To address this issue, Maragiannis et al. performed functional assessments in 3D printed models of severe AS [51,52]. Models of AS were fabricated from multiple patients, and functional parameters were tested under variable flow conditions through a pulsatile flow imaging circuit. The models were fabricated from CT imaging data with CAD software and multimaterial 3D printing. Rigid material (VeroWhitePlus RGD835, Objet) was utilized for calcified areas, and soft, rubber-like material (TangoPlus FLX930, Objet) was used for noncalcified areas. The 3D printed models were designed to replicate the specifics of diseased aortic valves including calcification, tissue thickening, and anatomic geometry (Fig. 6.4A). Proximal and distal aortic pressures were measured with pressure catheters, and ultrasound was used to determine stroke volumes and velocity time integrals. Doppler aortic valve area and mean gradient correlated very well with catheter-based valve area and gradient, and the difference between the clinical and 3D printed aortic valve area was small (Fig. 6.4B). Although the materials used for the 3D printed models were not the same as native biological tissue, the replicated degenerative aortic valves showed that fused dual-material designs were able to mimic the properties of native tissue. Such combined material anatomic modeling of aortic valves appears to have most value when periprocedural simulation is needed before aortic valve intervention. Anatomic and functional aortic valve modeling can optimize TAVR results by predicting complications, and definitively deserves further investigation. Variability of the aortic valve area at different blood flows (a paradigm of low-flow, low-gradient AS) suggests additional applications of 3D anatomic modeling when complex flow conditions arise from multiple cardiac structural pathologies [9].

One common complication of TAVR is heart block requiring cardiac pacemaker placement [53]. Ex vivo 3D printed model simulations have been used to investigate the mechanism responsible for permanent pacemaker implantation after TAVR [49]. The presence of preexisting right bundle branch block (RBBB) and heavy coronary calcium load of the left coronary cusp have been identified as the greatest predictors of permanent pacemaker placement after TAVR (Fig. 6.3). However, patients with preexisting RBBB and modest calcium load in the left coronary cusp are at a much lower risk for pacemaker implantation. Thus, development of conduction block and the need for pacemaker implantation may be related to asymmetric aortic calcium distribution. Ex vivo TAVR, including valve deployment and balloon inflation, can be simulated in 3D printed silicone aortic annuli. Such simulations revealed that the deployed valve and balloon inflation actually shifted calcium away from the left coronary cusp toward the right coronary cusp and noncoronary cusp commissure. These findings indicate that localized injury to conduction pathways adjacent to right and noncoronary cusps may be partially responsible for the development of heart block after TAVR.

Another common complication of surgical and transcatheter valve replacement is paravalvular leak [54]. Although the frequency of severe paravalvular leak has decreased due to improved valve size selection and positioning, mild and moderate

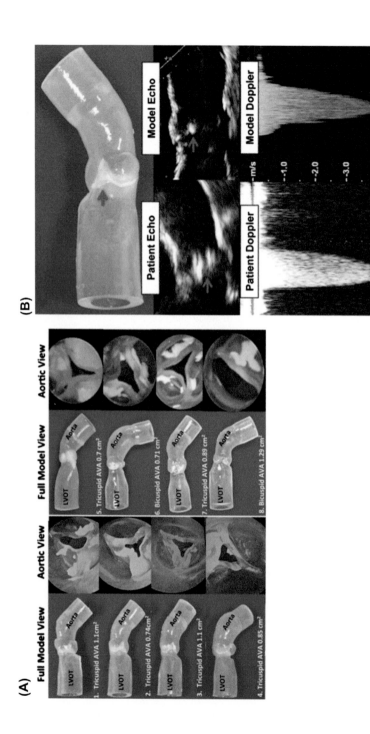

FIGURE 6.4 Patient-specific 3D models of the aortic valve.

(A) 3D printed models depicting the individual variation in calcium distribution patterns (*white*) and aortic valve orifice shape. *AVA*, aortic valve area; *LVOT*, left ventricular outflow tract. (B) Using echocardiography, the 3D printed model was shown to have good image quality with clear depiction of aortic valve leaflet calcification (*red arrow*). Doppler imaging also correlated well with the patient Doppler. Velocity and time scales are the same in both images.

Reproduced with permission from Maragiannis D, et al. Replicating patient-specific severe aortic valve stenosis with functional 3D modeling. Circ Cardiovasc Imaging 2015;8: e003626.

leaks are still common. Moderate paravalvular leaks may reduce functional benefit and survival after TAVR [30]. Anatomic modeling with 3D printing can help plan and achieve percutaneous repair of prosthetic aortic valve paravalvular leak [55]. Ripley et al. fabricated 3D printed models for TAVR planning. Light transmission testing was used for post-TAVR prediction of paravalvular AR. Valves were implanted in patient-specific aortic root models. Light was projected through the models as a surrogate for paravalvular leak. The presence and absence of paravalvular regurgitation was predicted in several patient models along with the location of regurgitation when it occurred, suggesting a role for 3D printing in the estimation of post-TAVR paravalvular leak [56].

While most investigations of 3D printing have focused on AS, 3D printed models have also been fabricated for AR. Functional flow models of AR have been examined with Doppler imaging and compared to their clinical counterparts [57]. Comparable parameters including peak velocity, regurgitant time, pressure half time, and regurgitant jets have been documented, thereby showing the potential of 3D printed models with applications for AR as well.

MITRAL VALVE DISEASE

MV disease is the most prevalent cardiac valve disorder. Its prevalence is >10% in elderly patients [22]. Mitral regurgitation (MR), by far the most common mitral pathology, is broadly classified as primary (a predominantly degenerative disease of the MV apparatus) or secondary (a disease of the LV, affecting the integrity and function of the MV apparatus) MR [58]. Mitral stenosis (MS) is commonly due to rheumatic disease, especially in the developing countries. However, senile calcification of the mitral annulus with extension into the leaflets can obstruct the LV inflow [27].

In primary MR, surgical repair is the standard of care and is associated with better outcomes than MV replacement [59]. The treatment of asymptomatic patients with severe MR in valve centers of excellence result in successful repair of >95% with surgical mortality of <1% [58]. However, since many patients are elderly with severe comorbidities and depressed LV function when referred for MR repair/replacement, almost 50% are denied surgery for primary MV [60]. As a result, with the increase in worldwide prevalence of degenerative MR, there is a considerable unmet clinical demand for a minimally invasive approach to MR [60]. Transcatheter MV repair with a MitraClip device may improve the outcome of patients with primary MR who are at high or prohibitive surgical risk [61,62].

In secondary MR, the rationale for surgical or transcatheter interventions is that lessening of MR may interrupt the downward spiral of volume overload causing more MR with the endpoint of improving survival. Various techniques of surgical MV repair have failed to improve survival in patients with secondary MR. Surgical MV repair has not been shown to improve survival in secondary MR [63]. Although it does not improve survival, valve sparing MV replacement is widely performed as it appears to be more durable than repair [63–65].

Routine use of sophisticated imaging modalities has led to better understanding of MV pathology and novel therapeutic approaches. Newer transcatheter approaches to MV repair/replacement are under investigation. They will clearly meet an unmet need in elderly patients with significant comorbidities, who have a high operative risk [58].

Current transcatheter therapy of the MV include the following: (1) the MitraClip leaflet repair system (Abbott Vascular, Abbott Park, IL); (2) early trials and ongoing development of transcatheter mitral valve replacement (TMVR) devices and repair systems; (3) the use of balloon-expandable transcatheter valves in valve-in-valve and valve-in-ring interventions; (4) and in some cases the use of closure devices to occlude a mitral leaflet perforation or paravalvular leaks [21]. Anatomic modeling with 3D printing is likely to play a major role in the selection and insertion of newer transcatheter devices for MV repair/replacement.

ANATOMY OF THE MITRAL VALVE

In contrast to the aortic valve, the MV apparatus comprises multiple complex structures that are essential for normal MV functioning [66]. As the MV is immediately adjacent to the aortic valve, its functioning is affected by aortic valve pathology [71] (Fig. 6.5). There are six components of the mitral apparatus, namely, the left atrial wall, annulus, leaflets, chordae tendinae, the two papillary muscles (PM), and LV wall.

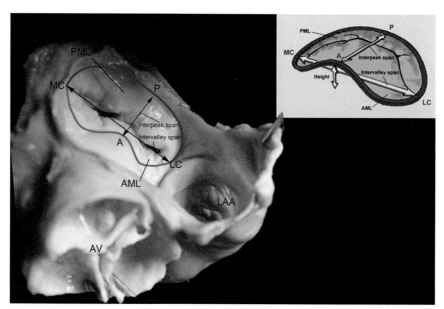

FIGURE 6.5 The saddle shape of the mitral valve annulus.

A, anterior; *AML*, anterior mitral leaflet; *AV*, aortic valve; *LAA*, left atrial appendage; *LC*, lateral commissure; *MC*, medial commissure; *P*, posterior; *PML*, posterior mitral leaflet.

Reproduced with permission from Perpetua EM, et al. Anatomy and function of the normal and diseased mitral apparatus: Implications for transcatheter therapy. Interv Cardiol Clin 2016;5:1–16.

Left Atrial Wall

Perloff and Roberts [67] emphasized the important role of the atrial wall in the finely tuned mitral mechanism. Left atrial enlargement per se can contribute to MR [67]. Because of the continuity of the atrial endocardium over the atrial surface of the posterior mitral leaflet, dilation of the left atrium (LA) exerts tension on the posterior leaflet making it vulnerable to being displaced [68].

Annulus

The MV is obliquely located in the heart and has a hyperbolic paraboloid shape similar to a saddle [66] (Fig. 6.5). The saddle peaks at the anterior and posterior horns and the nadirs are near the fibrous trigones. The anterior horn is continuous with the aortic valve anteriorly (forming the aorto-mitral continuity) and the posterior horn is formed by the insertion of the posterior mitral leaflet at the atrioventricular junction posteriorly [69]. The nadirs are defined as areas of collagenous thickening at the ends of the aorto-mitral curtain [69], which is part of the cardiac skeleton. While the term annulus implies a fibrous ring, it is only the anterior part of the annulus that is fibrous and rigid, making it less susceptible to dilatation [68].

When viewed en face, the mitral annulus is D-shaped, in which the anterior horn is the straight side and the posterior horn is curved (Fig. 6.6A). It is important to note that the annulus encroaches on the LVOT (Fig. 6.6B and C). Therefore, when sizing the annulus for TMVR, one must respect the anterior horn relationship to the LVOT; otherwise, inclusion of this component will lead to device encroachment into the LVOT and potentially fatal LVOT obstruction [69].

The posterior horn of the annulus is mostly muscular and lacks a well-formed fibrous cord [68]. The posterior annulus is therefore more vulnerable to degenerative changes, such as calcification and dilatation. With severe dilation, the minor axis of the annulus becomes distended, resulting in an oval-shaped orifice and compromising coaptation of the leaflets.

The mitral annulus is dynamic. During the cardiac cycle, the annulus area varies by 23%−40%. It is the lowest in midsystole, and the greatest during diastole [70]. Although annular shape and area strongly depend on LV systolic function, the nonplanar shape of the annulus remains somewhat preserved [71].

Leaflets

The MV has two leaflets: the anterior mitral leaflet (AML) and the posterior mitral leaflet (PML). The AML is tall, apical, and narrow, lining one-third of the annular circumference, while the PML is shorter and wider lining the remainder of the annular circumference. When viewing the MV en face in its closed state, the two leaflets can be seen forming an arc with the AML appearing to form the majority of the left atrial floor, although its area is approximately equal to that of the PML [68]. Each end of the closure line is referred to as a commissure: the anterolateral and posteromedial commissures. The PML has three scallops, lateral,

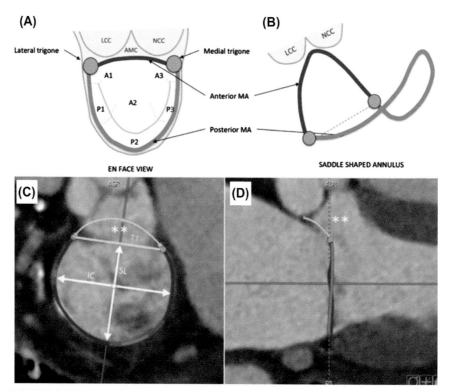

FIGURE 6.6 Mitral annular geometry and pertinent annular measurements for transcatheter device sizing.

Geometry of the mitral annulus and valve including anterior (A1-3) and posterior (P1-3) leaflets in relation to the lateral and medial fibrous trigones and aorto-mitral continuity (AMC) (A). The saddle-shaped annulus is shown in (B). For the purposes of transcatheter device sizing, 3D segmentation of the mitral annulus includes truncation of the anterior horn (*pink line*) to achieve the D-shaped mitral annulus to avoid encroachment of the left ventricular outflow tract (**) (C and D). *IC*, intercommissural distance; *LCC*, left coronary cusp; *NCC*, noncoronary cusp; *SL*, septal-lateral distance; *TT*, trigone-trigone distance.

Reproduced with permission from Naoum C, et al. Cardiac computed tomography and magnetic resonance imaging in the evaluation of mitral and tricuspid valve disease: Implications for transcatheter interventions. Circ Cardiovasc Imaging 2017;10 [pii:e005331].

middle, and medial, referred to as P1, P2, and P3, respectively. The AML does not have scallops but is correspondingly labeled A1, A2, and A3 (Fig. 6.6A).

Normal leaflets are thin and pliable. They can be affected by multiple primary and secondary causes of MV disease. In mitral valve prolapse (MVP), which is defined by leaflet extension into the LA beyond the plane of the atrioventricular junction (annular plane) during ventricular systole [68], the leaflets are floppy, thickened, and often gelatinous. In contrast, rheumatic heart disease results in stiffening of the leaflet tips and fusion of the commissures. Fibroelastic deficiency, frequently seen in the elderly, may result in less pliable leaflets and chordal rupture.

Chordae Tendinae

The tendinous cords are string-like structures that attach to the free edge of leaflets to the PM or directly to the posteroinferior ventricular wall. They are classified into three orders. The first order cords are numerous and delicate and are attached to the free edge leaflets. The second order thick cords attach to the ventricular surface of the leaflets beyond the free edge, forming the rough zone. Third order cords attach only to PML since they arise directly from the ventricular wall [66].

The Papillary Muscles and the Left Ventricular Wall

Each PM gives rise to chords that attach to both leaflets on its respective side, and insert into the ventricle at the middle and apical third [66]. The anterolateral PM has dual blood supply from the left coronary artery, while the posteromedial PM has single blood supply from the right coronary artery, making it more susceptible to ischemic injury.

3D PRINTING APPLICATIONS

The success of transcatheter therapy for MV disease heavily depends on accurate and patient-specific evaluation of MV apparatus anatomy and dysfunction. The mitral annular area, anterior mitral leaflet length, aorto-mitral angle, LVOT area, and mitral annular calcifications should be quantified [12]. Advanced cross-sectional imaging and echocardiography can clearly delineate most of the MV apparatus features. However, advanced cross-sectional imaging and echocardiography cannot predict patient-specific device-valve interaction. Anatomic modeling of the MV with 3D printing allows matching patient and device, thereby enhancing clinical outcome [18]. Binder et al. showed that stereolithographic modeling of TEE images of the MV could be done and provided a fairly accurate depiction of MV anatomy and pathology [72]. The feasibility, accuracy, and clinical potential of anatomic MV modeling with 3D printing were subsequently demonstrated [12,17−19,73].

Important anatomic and functional considerations of surgical or transcatheter therapies for MV disease are as follows:

Annular Size and Geometry

The orientation, shape, and dimensions of the MV annulus are extremely variable and independent of LV and LA size in normal subjects [69]. In patients with MR, annular dimensions are significantly larger, particularly in MVP, where annular geometry is distorted with septal-lateral expansion. Mitral annular area is 18% greater in patients with MVP than in patients with secondary MR [74]. In early TMVR experience, annular sizing was done using CT imaging. Nowadays, 3D TEE is used for D-shaped mitral annular segmentation without systematic differences in annular dimensions [74]. Routine TEE images provide accurate datasets for 3D printing of mitral annulus before and after repair surgeries [19]. Anatomic modeling of mitral annulus was found to closely match the geometry and size of normal and pathologic mitral annulus prior to surgery. Postsurgery anatomic modeling of the annulus showed close resemblance to the shape of the implanted ring [19] (Fig. 6.7). The advantage of 3D printed anatomic modeling is that sizing of the mitral annulus is based on geometric data dynamically acquired during the cardiac cycle. In contrast, visualization and sizing of the valve during surgery is performed on an arrested and physiologically unloaded heart [19]. Further studies will obviously be needed to show that anatomic modeling with 3D printing is superior, cost-effective, and provides longer lasting prosthetic annuli [75].

In secondary MR, atrioventricular remodeling deserves special attention as it affects the development and selection of TMVR devices. When the LV dilates, the normal linear relationship between the LA and LV myocardium comprises a posterior myocardial shelf [76,77] (Fig. 6.8C). The shelf size depends on many factors but is commonly larger in the setting of LV dilatation due to myocardial infarction. Of note, the shelf is dynamic with significant changes in size and configuration and may disappear in systole [69].

Fabricating anatomic models in systole and diastole helps to identify and size the shelf that provides capture and positioning for many of the investigational TMVR devices (Fig. 6.9). The posterior myocardial shelf observed in secondary MR, however, is typically not seen in primary MR patients who do not develop discordant LA and LV remodeling [69]. In MVP, the insertion of the mitral valve leaflet may be displaced into the LA resulting in the so-called mitral annular disjunction [78] (Fig. 6.8B).

Basal Left Ventricular Cavity

Evaluation of basal LV cavity size is important to ensure enough room for the TMVR device. Furthermore, size of PMs, presence of false bands, direct insertion of PM, and leaflet length need to be evaluated by CT. All these parameters affect selection of a TMVR device with adequate capture mechanism [69,77,79].

Pathologic Calcium Deposition

Mitral annular calcification (MAC) is a degenerative process that is associated with advancing age, and affects approximately 6% of the general population [80]. Severe calcifications can extend beyond the annulus and into the mitral leaflets, resulting in restricted leaflet motion and MS. The mechanism of MAC has been historically thought of as an age-related degenerative disorder. However, atherosclerosis-related calcification is related to MAC [81], in addition, to the abnormal calcium metabolism in chronic kidney disease [82].

CT is superior to 3D echocardiography for assessing location and extent of calcifications [69,83]. CT-derived 3D printing can accurately depict the size and location of calcium deposition and predict its interaction with devices [12,49] (Fig. 6.10). Little et al. reported a case of successful 3D anatomic modeling-

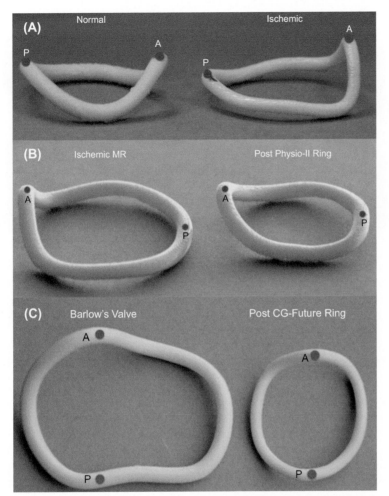

FIGURE 6.7 Echocardiography-derived 3D printed modeling of normal and abnormal mitral annuli.

(A) 3D printed annulus of a normal subject as compared to a patient with chronic significant ischemic mitral regurgitation. It can be seen that the ischemic annulus has lost the saddle shape and is dilated in the anterior-posterior axis as compared to the normal one. (B) 3D printed annulus of the same patient from (A) before and after undergoing repair with a Physio-II annuloplasty device (Edwards Life sciences, Irvine, CA, USA). (C) En-face view of a dilated mitral annulus of a patient with Barlow's degeneration of the mitral valve before and after repair with a Colvin-Galloway future ring (Medtronic, Minneapolis, MN, USA).

Reproduced with permission from Mahmood F, et al. Echocardiography derived three-dimensional printing of normal and abnormal mitral annuli. Ann Card Anaesth 2014;17:279–83.

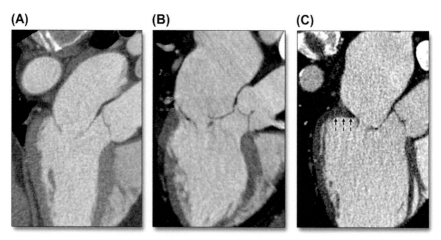

FIGURE 6.8 Geometry of the atrioventricular junction.

A normal atrioventricular junction is shown in a control subject (A). In a patient with mitral valve (MV) prolapse (B), there is a disjunction between the MV insertion point and the atrioventricular junction (*red arrows*). In a patient with functional mitral regurgitation (C), a prominent posterior myocardial shelf (*black arrows*) is appreciated, which is formed by the LV myocardium at the MV insertion point.

Reproduced with permission from Naoum C, et al. Mitral annular dimensions and geometry in patients with functional mitral regurgitation and mitral valve prolapse: Implications for transcatheter mitral valve implantation. JACC Cardiovasc Imaging 2016;9:269–80).

guided repair of symptomatic MR with restricted leaflet coaptation and a perforation of the posterior leaflet. Anatomic modeling with 3D printing of the MV apparatus was fabricated with CT datasets. The patient underwent MitraClip placement and an Amplatzer Duct Occluder (St. Jude Medical, St. Paul, Minnesota), which was sized by 3D anatomic modeling and showed to have minimal interaction with subvalvular calcification [21] (Fig. 6.11).

When homogenous and circumferential MAC can promote device anchoring for insertion of percutaneous aortic balloon expandable valves for MS [76]. However, extensive, nodular, and protruding calcium may result in device migration and significant paravalvular leak [84,85]. TMVR devices for severe MR have their own anchoring mechanisms now that are not reliant on the presence of MAC [76] (Fig. 6.9). Overall, when severe, MAC is considered a contraindication for TMVR in current feasibility studies [77].

Predicting LVOT Obstruction

LVOT area is decreased after mitral valve repair with annuloplasty rings or prostheses [86]. Following MV repair, systolic anterior motion of the MV (SAM) potentially leading to LVOT obstruction occurs in 2%–16% of patients. The main predictors of SAM and LVOT obstruction post-mitral valve repair are

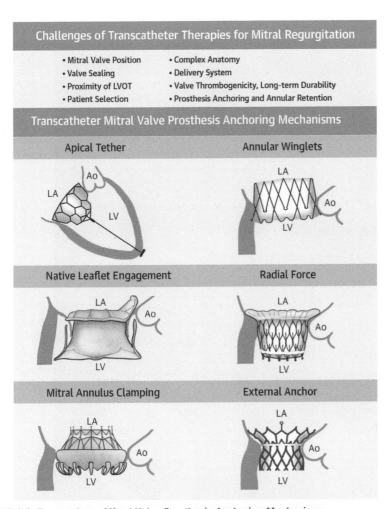

FIGURE 6.9 Transcatheter Mitral Valve Prosthesis Anchoring Mechanisms.

Reproduced with permission from Regueiro A, et al. Transcatheter mitral valve replacement: Insights from early clinical experience and future challenges. J Am Coll Cardiol 2017;69:2175–92.

as follows: (1) coaptation-to-septal distance <2.6 cm, (2) posterior mitral valve leaflet height >1.5 cm, (3) anterior leaflet/posterior leaflet height ratio of <1, and (4) aorto-mitral angle <130 degrees [87].

LVOT obstruction has been reported after surgical replacement. LVOT obstruction is usually due to incorrect strut position from a stented tissue valve [88,89], or from preserved mitral leaflet tissue [90]. Strut obstruction from mechanical mitral prostheses is less likely to cause LVOT obstruction and usually occurs due to malpositioning of the prosthesis [91].

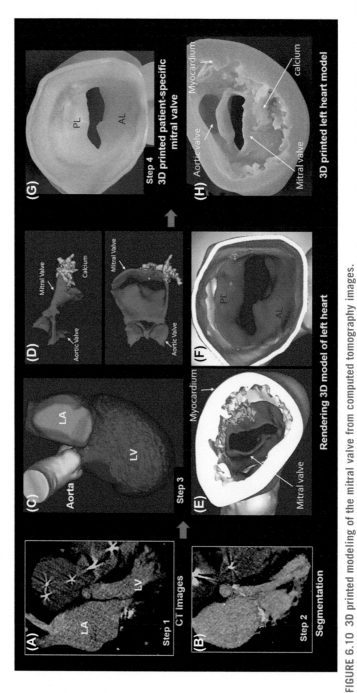

FIGURE 6.10 3D printed modeling of the mitral valve from computed tomography images.

Reproduced with permission from Vukicevic M, et al. 3D printed modeling of the mitral valve for catheter-based structural interventions. Ann Biomed Eng 2017;45:508−19.

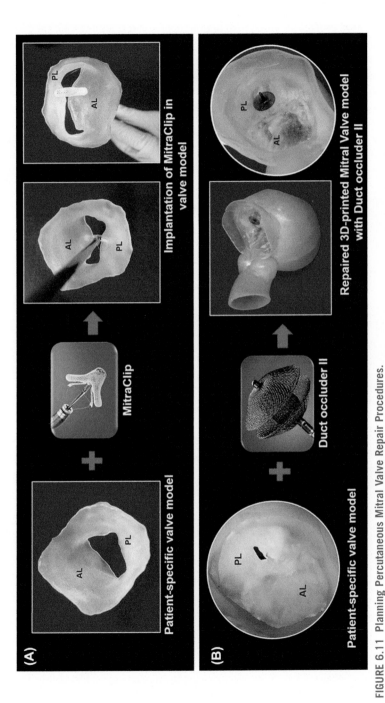

FIGURE 6.11 Planning Percutaneous Mitral Valve Repair Procedures.

Identifying LVOT obstruction by 2D echocardiography can be difficult in patients with a mitral prosthesis. It requires high temporal and spatial resolution imaging with MRI or CT. Both imaging techniques can accurately visualize preserved subvalvular apparatus and identify it as a cause of bio-prosthesis MV gradient.

As with surgical MV replacement or repair, LVOT obstruction can be a lethal complication of TMVR, mitral valve-in-valve, and valve-in-ring procedures, as well as implantation of transcatheter balloon-expandable valves in calcific MV disease [85,92]. The reported incidence of acute LVOT obstruction with transcatheter mitral valve-in-ring procedures is 8.2% and 9.3% after TMVR in patients with severe mitral annular calcification [93]. All these procedures ultimately lead to elongation of the outflow tract into the LV forming the so called "neo-LVOT" (Fig. 6.12),

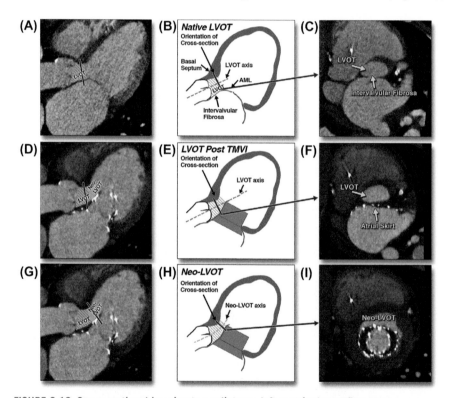

FIGURE 6.12 Cross-sectional imaging to predict neo-left ventricular outflow tract.

Native LVOT and neo-LVOT are demonstrated in systolic 3-chamber views using CT, schematic drawings, and cross-sectional short axis images. (A—C) Native LVOT is demonstrated. After transcatheter mitral valve replacement (Tendyne TMVR system, Tendyne Holdings, Roseville, Minnesota), there is no encroachment of the device onto the native LVOT cross-sectional area (D and E). TMVR resulting in elongation of the outflow tract into the left ventricle, creating the neo-LVOT (F—I).

Reproduced with permission from Blanke P, et al. Predicting LVOT obstruction in transcatheter mitral valve implantation: Concept of the neo-LVOT. JACC Cardiovasc Imaging 2017;10:482—5).

whereas the preexisting "native" LVOT confined by the most basal septum and the aorto-mitral curtain remains unchanged [92]. The mechanism of LVOT obstruction after device implantation is likely to result from fixed obstruction due to device encroachment into the LVOT or to SAM-induced dynamic LVOT obstruction [69].

Risk factors for narrow neo-LVOT include: (1) greater device protrusion into the LV; (2) greater device flaring at LVOT; (3) larger aorto-mitral angulation; and (4) more pronounced septal bulging [92] (Fig. 6.13). All of these factors can be assessed

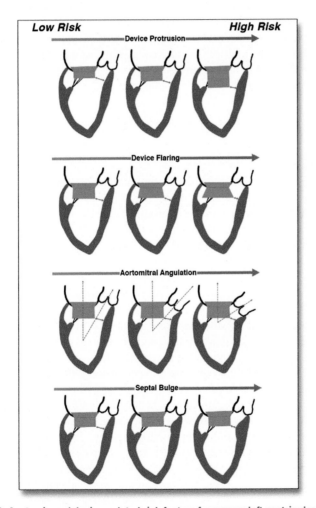

FIGURE 6.13 Anatomic and device-related risk factors for narrow left ventricular outflow tract.

Reproduced with permission from Blanke P, et al. Predicting LVOT obstruction in transcatheter mitral valve implantation: Concept of the neo-LVOT. JACC Cardiovasc Imaging 2017;10:482—5.

on preprocedural cross-sectional imaging, and patient-specific assessment can be performed using anatomic modeling and virtual device implantation. CT-derived accurate 3D anatomic modeling can predict LVOT obstruction and quantify neo-LVOT in TMVR [4] (Fig. 6.14). CT and CAD analysis can further help conceptualize the anatomy to test the effect of varying the above mentioned factors (angles, depths, and device flaring) of the implanted prosthesis.

One major challenge in MV anatomic modeling is the need for material blending, and material texture for replication of leaflets properties. A dual material using TangoPlus (Shore 27/Shore 35) approximates the stiffness of the so-called toe region (region of physiological MV function) of the stress-strain curve [12]. The combination of these materials mimics the tensile modulus of a normal porcine mitral valve leaflet [12]. Nevertheless, further testing of material blending is needed to replicate the multiple features of healthy and diseased MV complexes. Once done, the effect of device on anatomic configuration of native structures, and the effect of native structures on the implanted device, can be virtually and effectively tested.

TRICUSPID VALVE DISEASE

Tricuspid regurgitation (TR) is the most common right-sided valvular heart disease [94]. Trivial or mild TR is commonly observed and is well tolerated in normal individuals. However, significant and sustained TR is associated with volume overload and triggers progressive chamber dilation promoting atrial and ventricular arrhythmias. When left untreated, TR leads to increased morbidity and mortality [95].

Similar to MR, TR can be either primary, related to anatomic valvular problems, or secondary (functional) to right atrial (RA), right ventricular (RV), or both RA/RV remodeling. Secondary TR is most common in the Western world, and typically results from tricuspid annular dilation (secondary to RA or RV dilation) or because of tethering/tenting from RV dilation and papillary muscle displacement secondary to free wall dilation [69,94]. Significant functional TR may also occur in the context of left-sided heart disease with pulmonary hypertension [96].

Although functional TR responds to medical therapy, once annular dilatation occurs, tricuspid repair and replacement may be needed to prevent further progression and improve outcomes [97,98]. Patients are often referred for surgery late when the dilated RV fails. Surgical mortality for isolated tricuspid valve (TV) remains higher than for any other single valve surgery [99]. The negative impact of TR on survival of patients with left-sided valvular disease receiving transcatheter therapies [100,101] underscores the need for developing transcatheter therapy for TR. Currently, numerous transcatheter devices or early clinical trials are in development. They all require comprehensive and thorough understanding of TV anatomy and imaging.

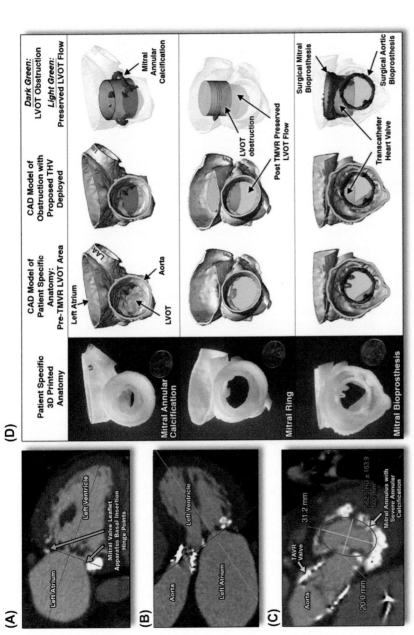

FIGURE 6.14 Computed tomography-guided sizing of the mitral annulus landing zone and 3D printed modeling to predict left ventricular outflow tract obstruction with transcatheter heart valves.

Crosshairs are aligned to the mitral annular plane in sagittal and coronal cross-sections (A and B) allowing for mitral annular dimensions to be obtained on axial thin sections (C). Computer-aided design (CAD)-generated valves are virtually tested on-screen and physically in the patient's 3D printed anatomy.

Once the proposed transcatheter heart valve size was identified, computer-aided design-generated Sapien, Sapien XT (SXT) and Sapien 3 valves (Edwards Lifesciences, Irvine, California), were virtually tested on screen and again physically in the patient's 3D printed mitral valve model to confirm sizing and predict LVOT obstruction.

Reproduced with permission from Wang DD, et al. Predicting LVOT obstruction after TMVR. JACC Cardiovasc Imaging 2016;9:1349–52.

ANATOMY OF THE TRICUSPID VALVE

Like the MV, the TV apparatus is complex, consisting of four major components: the three leaflets (anterior, posterior, and septal), chordae tendinae, three discrete papillary muscles (septal, posterior, and anterior), and the fibrous tricuspid annulus (with the attached atrium and ventricle).

There is a large variability in the anatomic features of each of the three leaflets. The relative circumferential or annular ratios of the anterior:septal:posterior leaflets in normal subjects are 1:1:0.75 [95]. The anterior leaflet is the largest and the most mobile, while the septal leaflet is the shortest in the radial direction and the least mobile. The septal leaflet originates medially from the tricuspid annulus itself directly above the interventricular septum. The posterior leaflet usually displays multiple scallops and is the shortest circumferentially [95].

The tricuspid annulus is much larger than the mitral annulus. The tricuspid annulus is a complex and dynamic structure that is affected by loading conditions [69]. Unlike the mitral valve, there is no fibrous continuity with the corresponding semilunar valve [95]. A normal annulus is ovoid and saddle shaped, displaced towards the atrium in the antero-septal portion near the RV outflow tract (RVOT) and the posterolateral portion, and displaced towards the apex in the postero-septal portion near the inflow of the coronary sinus and the anterolateral segment [102]. The tricuspid annulus has nonplanar, nonsingle plane structure by 3D echocardiography in healthy subjects and a dilated (mostly in the septal-lateral direction) planar structure in patients with functional TR [103]. Design of a tricuspid ring, according to 3D imaging data, may improve atrioventricular dynamics and reduce leaflet stress [103].

The TV apparatus includes two distinct papillary muscles (anterior and posterior) and a third variable papillary muscle. The anterior papillary muscle is typically the largest with chordal attachments supporting the anterior and posterior leaflets. The posterior papillary muscle is often bifid or trifid, lending chordal support to the posterior and septal leaflets. The septal papillary muscle is variable and may be absent in up to 20% of normal subjects. Unlike the mitral valve, chordae may arise directly from the septum to the anterior and septal leaflets [95]. These attachments are partly responsible for high prevalence of TR whenever the RV dilates.

Many of the above mentioned anatomic issues represent a challenge for the surgical and transcatheter treatment of TR. The nonplanar structure of the tricuspid annulus, lack of calcium (compared with mitral or aortic valve), and the dimension of the tricuspid annulus (>40 mm) are the main challenges when considering transcatheter TV replacement. Moreover, RV morphology and proximity to other structures, such as the coronary sinus, the right coronary artery, AV node, and bundle of His, increases the risk of complications [102].

3D PRINTING APPLICATIONS

Echocardiographic imaging of the TV plays an essential role in quantifying TR and chamber size and function [27], as well as guiding transcatheter procedures [95]. Besides echocardiography, CMR provides useful information regarding the hemodynamic severity of TV disease and extent of RV remodeling, while CT provides important morphological data [69]. All of these imaging modalities can provide accurate volumetric datasets for 3D printing.

Anatomic modeling of the TV with 3D printing can be fabricated from 3D TTE with accuracy of <1 mm valvular dimension difference between modeling and echocardiographic data [104]. In porcine models, 3D modeling of the right-sided cardiac chambers was fabricated from CT data to help in testing compressible nitinol stents for biological valve prostheses [105]. Additionally, CT data from human anatomy have been used for 3D anatomic modeling to plan and simulate transcatheter tricuspid procedures. Anatomic modeling with 3D printing is helpful to train operators for transcatheter TV procedures without the need for animal testing [106].

Anatomic modeling with 3D printing was reported in a transcatheter caval tricuspid valve implantation (CAVI) procedure [107]. The patient had been referred for CAVI due to prohibitive surgical risk. Preprocedure CT imaging of the chest and abdomen were completed, and anatomic modeling of the right atrium-inferior vena cava was fabricated to assist with transcatheter valve selection and implantation (Fig. 6.15). The feasibility and utility of anatomic modeling with 3D printing for TV intervention need to be further assessed.

PULMONARY VALVE DISEASE

Diseases of the pulmonary valve are often associated with congenital heart disease, and significant lesions are managed with surgical or transcatheter valve replacement [1,30]. However, long-term dysfunction of the RVOT with pulmonary stenosis or regurgitation after repair of congenital heart defects is common. Pulmonary valve disease can be managed by transcatheter approach in order to avoid additional open-heart surgery and reduce procedural risks. However, variable characteristics of the RVOT can render difficult transcatheter pulmonary valve procedure in patients with adult congenital heart disease [108–110].

Preprocedural modeling is often performed with CMR imaging, although it has not been proven to improve procedural success. Schievano and colleagues have examined whether 3D printed models of the RVOT and pulmonary artery based on MRI are useful for selecting candidates for percutaneous pulmonary valve intervention [110]. These investigators created 3D models from MRI data, which were then reviewed retrospectively, to assess whether the percutaneous valve intervention should have been performed. They concluded that 3D rapid modeling provided more information than CMR imaging when selecting patients for a percutaneous intervention. Rapid 3D anatomic modeling from 3D echocardiographic datasets is currently being investigated to assist in patient selection, prosthesis preparation, and percutaneous pulmonary valve implantation [111].

Many patients with tetralogy of Fallot who develop pulmonary valve disease after repair are not candidates for transcatheter pulmonary valve replacement because

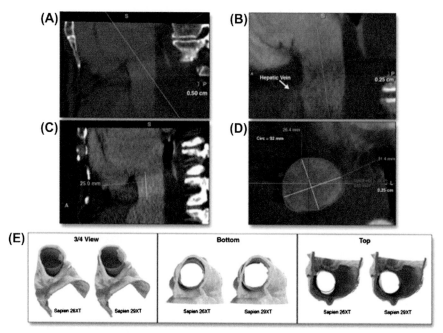

FIGURE 6.15 CT and 3D print guided selection of transcatheter valve for caval valve implantation.

Measurement of the distance from the right atrium—inferior vena cava (RA-IVC) junction plane (A) to the superior level of the first hepatic vein (B and C) was taken to aid in optimum positioning of the valve. Then, measurements of the corresponding region in short axis were taken (D). 3D printed models of the RA-IVC junction were made with SAPIEN 26XT and 29XT (Edwards Lifesciences Corp., Irvine, California) valve models inserted for fit testing. Large gaps between the IVC wall and the SAPIEN 26XT valve frame were demonstrated, whereas SAPIEN 29XT showed a good fit and was chosen from the procedure (E).

Reproduced with permission from O'Neill B, et al. Transcatheter caval valve implantation using multimodality imaging: roles of TEE, CT, and 3D printing. JACC Cardiovasc Imaging 2015;8:221–5.

of large and irregular RVOT [112]. Phillips et al. reported their experience with 3D RVOT modeling: Eight patients underwent 3D printed RVOT modeling in order to simulate geometric reforming with a prestented landing zone of the RVOT before transcatheter valvular implantation. Transcatheter pulmonary valves were successfully implanted in all patients without significant complications and with only trivial pulmonary regurgitation at follow-up. For this complex procedure, 3D printed models have been shown to be useful in preprocedural planning, and others have performed similar repair in large RVOT [113].

DEVELOPMENT OF PERSONALIZED VALVULAR INTERVENTIONS

As medical and surgical treatments for valvular heart disease continue to evolve, tissue valves that can adapt to a changing biological environment are likely to be developed. Substantial research has been undertaken within the realm of tissue engineering to create functional, biologically living heart valves [79]. Engineering strategies include molded scaffolds that use both naturally derived and synthetic polymers along with 3D biological modeling. However, there is currently no living viable material truly capable of biological growth and integration into a patient's native tissue. Stem cell or other cellular engineering research may help fill the present gap.

In the meantime several issues regarding heart valve anatomic modeling with 3D printing must be addressed. First, the most accurate imaging strategy to build well-validated 3D valvular modeling must be determined. All aspects of modeling need to be optimized including image acquisition, segmentation, 3D image reconstruction, digital software processing, and multimaterial for 3D modeling. Next, upon accurate model creation, the hemodynamics of valvular models must be tested and validated against functional human anatomy. Additionally, the combination of materials used for fabricating 3D models deserves further study to develop the most accurate replication of human cardiac valve disease. The hope is that materials similar to native human tissues will be developed so that anatomic and functional modeling with 3D printing will allow design and testing of biologically adaptable and patient-specific cardiac valve prosthesis.

REFERENCES

[1] Giannopoulos AA, Mitsouras D, Yoo SJ, Liu PP, Chatzizisis YS, Rybicki FJ. Applications of 3D printing in cardiovascular diseases. Nat Rev Cardiol December 2016; 13(12):701−18.

[2] Gallo M, D'Onofrio A, Tarantini G, Nocerino E, Remondino F, Gerosa G. 3D-printing model for complex aortic transcatheter valve treatment. Int J Cardiol May 1, 2016;210: 139−40.

[3] Schmauss D, Schmitz C, Bigdeli AK, Weber S, Gerber N, Beiras-Fernandez A, et al. Three-dimensional printing of models for preoperative planning and simulation of transcatheter valve replacement. Ann Thorac Surg February 2012;93(2):e31−3.

[4] Wang DD, Eng M, Greenbaum A, Myers E, Forbes M, Pantelic M, et al. Predicting LVOT obstruction after TMVR. JACC Cardiovasc Imaging November 2016;9(11): 1349−52.

[5] Lim KH, Loo ZY, Goldie SJ, Adams JW, McMenamin PG. Use of 3D printed models in medical education: a randomized control trial comparing 3D prints versus cadaveric materials for learning external cardiac anatomy. Anat Sci Educ May 6, 2016;9(3): 213−21.

[6] Costello JP, Olivieri LJ, Su L, Krieger A, Alfares F, Thabit O, et al. Incorporating three-dimensional printing into a simulation-based congenital heart disease and critical care training curriculum for resident physicians. Congenit Heart Dis March—April 2015;10(2):185—90.

[7] Olivieri LJ, Su L, Hynes CF, Krieger A, Alfares FA, Ramakrishnan K, et al. "Just-in-time" simulation training using 3-D printed cardiac models after congenital cardiac surgery. World J Pediatr Congenit Heart Surg March 2016;7(2):164—8.

[8] Giannopoulos AA, Steigner ML, George E, Barile M, Hunsaker AR, Rybicki FJ, et al. Cardiothoracic applications of 3-dimensional printing. J Thorac Imaging September 2016;31(5):253—72.

[9] Vukicevic M, Mosadegh B, Min JK, Little SH. Cardiac 3D printing and its future directions. JACC Cardiovasc Imaging February 2017;10(2):171—84.

[10] Mashari A, Knio Z, Jeganathan J, Montealegre-Gallegos M, Yeh L, Amador Y, et al. Hemodynamic testing of patient-specific mitral valves using a pulse duplicator: a clinical application of three-dimensional printing. J Cardiothorac Vasc Anesth October 2016;30(5):1278—85.

[11] Kurup HK, Samuel BP, Vettukattil JJ. Hybrid 3D printing: a game-changer in personalized cardiac medicine? Expert Rev Cardiovasc Ther December 2015;13(12): 1281—4.

[12] Vukicevic M, Puperi DS, Jane Grande-Allen K, Little SH. 3D printed modeling of the mitral valve for catheter-based structural interventions. Ann Biomed Eng February 2017;45(2):508—19.

[13] Gosnell J, Pietila T, Samuel BP, Kurup HK, Haw MP, Vettukattil JJ. Integration of computed tomography and three-dimensional echocardiography for hybrid three-dimensional printing in congenital heart disease. J Digit Imaging December 2016; 29(6):665—9.

[14] Meier LM, Meineri M, Qua Hiansen J, Horlick EM. Structural and congenital heart disease interventions: the role of three-dimensional printing. Neth Heart J February 2017;25(2):65—75.

[15] Mathur M, Patil P, Bove A. The role of 3D printing in structural heart disease: all that glitters is not gold. JACC Cardiovasc Imaging August 2015;8(8):987—8.

[16] Byrne N, Velasco Forte M, Tandon A, Valverde I, Hussain T. A systematic review of image segmentation methodology, used in the additive manufacture of patient-specific 3D printed models of the cardiovascular system. JRSM Cardiovasc Dis April 29, 2016;5. 2048004016645467.

[17] Mahmood F, Owais K, Taylor C, Montealegre-Gallegos M, Manning W, Matyal R, et al. Three-dimensional printing of mitral valve using echocardiographic data. JACC Cardiovasc Imaging February 2015;8(2):227—9.

[18] Witschey WR, Pouch AM, McGarvey JR, Ikeuchi K, Contijoch F, Levack MM, et al. Three-dimensional ultrasound-derived physical mitral valve modeling. Ann Thorac Surg August 2014;98(2):691—4.

[19] Mahmood F, Owais K, Montealegre-Gallegos M, Matyal R, Panzica P, Maslow A, et al. Echocardiography derived three-dimensional printing of normal and abnormal mitral annuli. Ann Card Anaesth October—December 2014;17(4):279—83.

[20] Markl M, Schumacher R, Kuffer J, Bley TA, Hennig J. Rapid vessel prototyping: vascular modeling using 3t magnetic resonance angiography and rapid prototyping technology. MAGMA December 2005;18(6):288—92.

[21] Little SH, Vukicevic M, Avenatti E, Ramchandani M, Barker CM. 3D printed modeling for patient-specific mitral valve intervention: repair with a clip and a plug. JACC Cardiovasc Interv May 9, 2016;9(9):973−5.

[22] Nkomo VT, Gardin JM, Skelton TN, Gottdiener JS, Scott CG, Enriquez-Sarano M. Burden of valvular heart diseases: a population-based study. Lancet September 16, 2006;368(9540):1005−11.

[23] Otto CM, Prendergast B. Aortic-valve stenosis−from patients at risk to severe valve obstruction. N Engl J Med August 21, 2014;371(8):744−56.

[24] Siu SC, Silversides CK. Bicuspid aortic valve disease. J Am Coll Cardiol June 22, 2010;55(25):2789−800.

[25] Kapadia SR, Tuzcu EM, Makkar RR, Svensson LG, Agarwal S, Kodali S, et al. Long-term outcomes of inoperable patients with aortic stenosis randomly assigned to transcatheter aortic valve replacement or standard therapy. Circulation October 21, 2014;130(17):1483−92.

[26] Makkar RR, Fontana GP, Jilaihawi H, Kapadia S, Pichard AD, Douglas PS, et al. Transcatheter aortic-valve replacement for inoperable severe aortic stenosis. N Engl J Med May 3, 2012;366(18):1696−704.

[27] Nishimura RA, Otto CM, Bonow RO, Carabello BA, Erwin 3rd JP, Guyton RA, et al. 2014 AHA/ACC guideline for the management of patients with valvular heart disease: a report of the American College of Cardiology/American Heart Association Task Force on Practice Guidelines. J Am Coll Cardiol June 10, 2014;63(22):e57−185.

[28] Nishimura RA, Otto CM, Bonow RO, Carabello BA, Erwin 3rd JP, Fleisher LA, et al. 2017 AHA/ACC focused update of the 2014 AHA/ACC guideline for the management of patients with valvular heart disease: a report of the American College of Cardiology/American Heart Association Task Force on Clinical Practice Guidelines. J Am Coll Cardiol July 11, 2017;70(2):252−89.

[29] Moat NE. Will TAVR become the predominant method for treating severe aortic stenosis? N Engl J Med April 28, 2016;374(17):1682−3.

[30] Figulla HR, Webb JG, Lauten A, Feldman T. The transcatheter valve technology pipeline for treatment of adult valvular heart disease. Eur Heart J July 21, 2016; 37(28):2226−39.

[31] Leon MB, Smith CR, Mack MJ, Makkar RR, Svensson LG, Kodali SK, et al. Transcatheter or surgical aortic-valve replacement in intermediate-risk patients. N Engl J Med April 28, 2016;374(17):1609−20.

[32] Sawaya FJ, Deutsch MA, Seiffert M, Yoon SH, Codner P, Wickramarachchi U, et al. Safety and efficacy of transcatheter aortic valve replacement in the treatment of pure aortic regurgitation in native valves and failing surgical bioprostheses: results from an International Registry study. JACC Cardiovasc Interv May 22, 2017;10(10):1048−56.

[33] Anderson RH. Clinical anatomy of the aortic root. Heart December 2000;84(6): 670−3.

[34] Piazza N, de Jaegere P, Schultz C, Becker AE, Serruys PW, Anderson RH. Anatomy of the aortic valvar complex and its implications for transcatheter implantation of the aortic valve. Circ Cardiovasc Interv August 2008;1(1):74−81.

[35] Loukas M, Bilinsky E, Bilinsky S, Blaak C, Tubbs RS, Anderson RH. The anatomy of the aortic root. Clin Anat July 2014;27(5):748−56.

[36] Brewer RJ, Deck JD, Capati B, Nolan SP. The dynamic aortic root. Its role in aortic valve function. J Thorac Cardiovasc Surg September 1976;72(3):413−7.

[37] Crawford MH, Roldan CA. Prevalence of aortic root dilatation and small aortic roots in valvular aortic stenosis. Am J Cardiol June 1, 2001;87(11):1311–3.

[38] Schaefer BM, Lewin MB, Stout KK, Gill E, Prueitt A, Byers PH, et al. The bicuspid aortic valve: an integrated phenotypic classification of leaflet morphology and aortic root shape. Heart December 2008;94(12):1634–8.

[39] Kang JW, Song HG, Yang DH, Baek S, Kim DH, Song JM, et al. Association between bicuspid aortic valve phenotype and patterns of valvular dysfunction and bicuspid aortopathy: comprehensive evaluation using MDCT and echocardiography. JACC Cardiovasc Imaging February 2013;6(2):150–61.

[40] Detaint D, Michelena HI, Nkomo VT, Vahanian A, Jondeau G, Sarano ME. Aortic dilatation patterns and rates in adults with bicuspid aortic valves: a comparative study with Marfan syndrome and degenerative aortopathy. Heart January 2014;100(2):126–34.

[41] Thubrikar M, Piepgrass WC, Shaner TW, Nolan SP. The design of the normal aortic valve. Am J Physiol December 1981;241(6):H795–801.

[42] Akins CW, Miller DC, Turina MI, Kouchoukos NT, Blackstone EH, Grunkemeier GL, et al. Guidelines for reporting mortality and morbidity after cardiac valve interventions. J Thorac Cardiovasc Surg April 2008;135(4):732–8.

[43] Kasel AM, Cassese S, Bleiziffer S, Amaki M, Hahn RT, Kastrati A, et al. Standardized imaging for aortic annular sizing: implications for transcatheter valve selection. JACC Cardiovasc Imaging February 2013;6(2):249–62.

[44] Tops LF, Wood DA, Delgado V, Schuijf JD, Mayo JR, Pasupati S, et al. Noninvasive evaluation of the aortic root with multislice computed tomography implications for transcatheter aortic valve replacement. JACC Cardiovasc Imaging May 2008;1(3):321–30.

[45] Webb JG, Chandavimol M, Thompson CR, Ricci DR, Carere RG, Munt BI, et al. Percutaneous aortic valve implantation retrograde from the femoral artery. Circulation February 14, 2006;113(6):842–50.

[46] Dashkevich A, Blanke P, Siepe M, Pache G, Langer M, Schlensak C, et al. Preoperative assessment of aortic annulus dimensions: comparison of noninvasive and intraoperative measurement. Ann Thorac Surg March 2011;91(3):709–14.

[47] Wang H, Hanna JM, Ganapathi A, Keenan JE, Hurwitz LM, Vavalle JP, et al. Comparison of aortic annulus size by transesophageal echocardiography and computed tomography angiography with direct surgical measurement. Am J Cardiol June 1, 2015;115(11):1568–73.

[48] Fujita B, Kutting M, Scholtz S, Utzenrath M, Hakim-Meibodi K, Paluszkiewicz L, et al. Development of an algorithm to plan and simulate a new interventional procedure. Interact Cardiovasc Thorac Surg July 2015;21(1):87–95.

[49] Fujita B, Kutting M, Seiffert M, Scholtz S, Egron S, Prashovikj E, et al. Calcium distribution patterns of the aortic valve as a risk factor for the need of permanent pacemaker implantation after transcatheter aortic valve implantation. Eur Heart J Cardiovasc Imaging December 2016;17(12):1385–93.

[50] Qian Z, Wang K, Chang Y, Zhang C, Wang B, Rajagopal V, et al. 3-D printing of biological tissue-mimicking aortic root using a novel meta-material technique: potential clinical applications. J Am Coll Cardiol April 2016;67(13):7.

[51] D JMS, Igo SR, Chang SM, Zoghbi WA, Little SH. Functional 3D printed patient-specific modeling of severe aortic stenosis. J Am Coll Cardiol September 9, 2014;64(10):1066–8.

[52] Maragiannis D, Jackson MS, Igo SR, Schutt RC, Connell P, Grande-Allen J, et al. Replicating patient-specific severe aortic valve stenosis with functional 3D modeling. Circ Cardiovasc Imaging October 2015;8(10):e003626.

[53] Khatri PJ, Webb JG, Rodes-Cabau J, Fremes SE, Ruel M, Lau K, et al. Adverse effects associated with transcatheter aortic valve implantation: a meta-analysis of contemporary studies. Ann Intern Med January 1, 2013;158(1):35−46.

[54] Leon MB, Smith CR, Mack M, Miller DC, Moses JW, Svensson LG, et al. Transcatheter aortic-valve implantation for aortic stenosis in patients who cannot undergo surgery. N Engl J Med October 21, 2010;363(17):1597−607.

[55] Sorajja P, Cabalka AK, Hagler DJ, Rihal CS. Long-term follow-up of percutaneous repair of paravalvular prosthetic regurgitation. J Am Coll Cardiol November 15, 2011;58(21):2218−24.

[56] Ripley B, Kelil T, Cheezum MK, Goncalves A, Di Carli MF, Rybicki FJ, et al. 3D printing based on cardiac CT assists anatomic visualization prior to transcatheter aortic valve replacement. J Cardiovasc Comput Tomogr January−February 2016;10(1): 28−36.

[57] Vukicevic M, Maragiannis D, Jackson M, Little SH. Abstract 18647: functional evaluation of a patient-specific 3D printed model of aortic regurgitation. Lippincott Williams & Wilkins Circulation 2015;132(Suppl 3):A18647.

[58] Nishimura RA, Vahanian A, Eleid MF, Mack MJ. Mitral valve disease—current management and future challenges. Lancet March 26, 2016;387(10025):1324−34.

[59] Enriquez-Sarano M, Schaff HV, Orszulak TA, Tajik AJ, Bailey KR, Frye RL. Valve repair improves the outcome of surgery for mitral regurgitation. A multivariate analysis. Circulation February 15, 1995;91(4):1022−8.

[60] Mirabel M, Iung B, Baron G, Messika-Zeitoun D, Detaint D, Vanoverschelde JL, et al. What are the characteristics of patients with severe, symptomatic, mitral regurgitation who are denied surgery? Eur Heart J June 2007;28(11):1358−65.

[61] Velazquez EJ, Samad Z, Al-Khalidi HR, Sangli C, Grayburn PA, Massaro JM, et al. The MitraClip and survival in patients with mitral regurgitation at high risk for surgery: a propensity-matched comparison. Am Heart J November 2015;170(5): 1050−1059.e3.

[62] Van den Branden BJ, Swaans MJ, Post MC, Rensing BJ, Eefting FD, Jaarsma W, et al. Percutaneous edge-to-edge mitral valve repair in high-surgical-risk patients: do we hit the target? JACC Cardiovasc Interv January 2012;5(1):105−11.

[63] Magne J, Girerd N, Senechal M, Mathieu P, Dagenais F, Dumesnil JG, et al. Mitral repair versus replacement for ischemic mitral regurgitation: comparison of short-term and long-term survival. Circulation September 15, 2009;120(11 Suppl): S104−11.

[64] Acker MA, Parides MK, Perrault LP, Moskowitz AJ, Gelijns AC, Voisine P, et al. Mitral-valve repair versus replacement for severe ischemic mitral regurgitation. N Engl J Med January 2, 2014;370(1):23−32.

[65] Goldstein D, Moskowitz AJ, Gelijns AC, Ailawadi G, Parides MK, Perrault LP, et al. Two-year outcomes of surgical treatment of severe ischemic mitral regurgitation. N Engl J Med January 28, 2016;374(4):344−53.

[66] Perpetua EM, Levin DB, Reisman M. Anatomy and function of the normal and diseased mitral apparatus: implications for transcatheter therapy. Interv Cardiol Clin January 2016;5(1):1−16.

[67] Perloff JK, Roberts WC. The mitral apparatus. Functional anatomy of mitral regurgitation. Circulation August 1972;46(2):227−39.

[68] Ho SY. Anatomy of the mitral valve. Heart November 2002;88(Suppl 4):iv5−10.

[69] Naoum C, Blanke P, Cavalcante JL, Leipsic J. Cardiac computed tomography and magnetic resonance imaging in the evaluation of mitral and tricuspid valve disease: implications for transcatheter interventions. Circ Cardiovasc Imaging March 2017; 10(3). https://doi.org/10.1161/CIRCIMAGING.116.005331.

[70] Ormiston JA, Shah PM, Tei C, Wong M. Size and motion of the mitral valve annulus in man. I. A two-dimensional echocardiographic method and findings in normal subjects. Circulation July 1981;64(1):113−20.

[71] Flachskampf FA, Chandra S, Gaddipatti A, Levine RA, Weyman AE, Ameling W, et al. Analysis of shape and motion of the mitral annulus in subjects with and without cardiomyopathy by echocardiographic 3-dimensional reconstruction. J Am Soc Echocardiogr April 2000;13(4):277−87.

[72] Binder TM, Moertl D, Mundigler G, Rehak G, Franke M, Delle-Karth G, et al. Stereolithographic biomodeling to create tangible hard copies of cardiac structures from echocardiographic data: in vitro and in vivo validation. J Am Coll Cardiol January 2000;35(1):230−7.

[73] Owais K, Pal A, Matyal R, Montealegre-Gallegos M, Khabbaz KR, Maslow A, et al. Three-dimensional printing of the mitral annulus using echocardiographic data: science fiction or in the operating room next door? J Cardiothorac Vasc Anesth October 2014;28(5):1393−6.

[74] Mak GJ, Blanke P, Ong K, Naoum C, Thompson CR, Webb JG, et al. Three-dimensional echocardiography compared with computed tomography to determine mitral annulus size before transcatheter mitral valve implantation. Circ Cardiovasc Imaging June 2016;9(6). https://doi.org/10.1161/CIRCIMAGING.115.004176.

[75] Kapur KK, Garg N. Echocardiography derived three- dimensional printing of normal and abnormal mitral annuli. Ann Card Anaesth October−December 2014;17(4): 283−4.

[76] Naoum C, Leipsic J, Cheung A, Ye J, Bilbey N, Mak G, et al. Mitral annular dimensions and geometry in patients with functional mitral regurgitation and mitral valve prolapse: implications for transcatheter mitral valve implantation. JACC Cardiovasc Imaging March 2016;9(3):269−80.

[77] Blanke P, Naoum C, Webb J, Dvir D, Hahn RT, Grayburn P, et al. Multimodality imaging in the context of transcatheter mitral valve replacement: establishing consensus among modalities and disciplines. JACC Cardiovasc Imaging October 2015;8(10):1191−208.

[78] Hutchins GM, Moore GW, Skoog DK. The association of floppy mitral valve with disjunction of the mitral annulus fibrosus. N Engl J Med February 27, 1986;314(9): 535−40.

[79] Cheung DY, Duan B, Butcher JT. Current progress in tissue engineering of heart valves: multiscale problems, multiscale solutions. Expert Opin Biol Ther 2015; 15(8):1155−72.

[80] Fox CS, Vasan RS, Parise H, Levy D, O'Donnell CJ, D'Agostino RB, et al. Mitral annular calcification predicts cardiovascular morbidity and mortality: the Framingham Heart Study. Circulation March 25, 2003;107(11):1492−6.

[81] Allison MA, Cheung P, Criqui MH, Langer RD, Wright CM. Mitral and aortic annular calcification are highly associated with systemic calcified atherosclerosis. Circulation February 14, 2006;113(6):861−6.

[82] Abd Alamir M, Radulescu V, Goyfman M, Mohler 3rd ER, Gao YL, Budoff MJ, et al. Prevalence and correlates of mitral annular calcification in adults with chronic kidney disease: results from CRIC study. Atherosclerosis September 2015;242(1):117−22.

[83] Mejean S, Bouvier E, Bataille V, Seknadji P, Fourchy D, Tabet JY, et al. Mitral annular calcium and mitral stenosis determined by multidetector computed tomography in patients referred for aortic stenosis. Am J Cardiol October 15, 2016;118(8):1251−7.

[84] Nguyen A, Urena M, Himbert D, Goublaire C, Brochet E, Gardy-Verdonk C, et al. Late displacement after transcatheter mitral valve replacement for degenerative mitral valve disease with massive annular calcification. JACC Cardiovasc Interv August 8, 2016; 9(15):1633−4.

[85] Bapat VV, Khaliel F, Ihleberg L. Delayed migration of Sapien valve following a transcatheter mitral valve-in-valve implantation. Catheter Cardiovasc Interv January 1, 2014;83(1):E150−4.

[86] Lee KS, Stewart WJ, Lever HM, Underwood PL, Cosgrove DM. Mechanism of outflow tract obstruction causing failed mitral valve repair. Anterior displacement of leaflet coaptation. Circulation November 1993;88(5 Pt 2):II24−I29.

[87] Maslow AD, Regan MM, Haering JM, Johnson RG, Levine RA. Echocardiographic predictors of left ventricular outflow tract obstruction and systolic anterior motion of the mitral valve after mitral valve reconstruction for myxomatous valve disease. J Am Coll Cardiol December 1999;34(7):2096−104.

[88] Guler N, Ozkara C, Akyol A. Left ventricular outflow tract obstruction after bioprosthetic mitral valve replacement with posterior mitral leaflet preservation. Tex Heart Inst J 2006;33(3):399−401.

[89] Jett GK, Jett MD, Bosco P, van Rijk-Swikker GL, Clark RE. Left ventricular outflow tract obstruction following mitral valve replacement: effect of strut height and orientation. Ann Thorac Surg September 1986;42(3):299−303.

[90] Wu Q, Zhang L, Zhu R. Obstruction of left ventricular outflow tract after mechanical mitral valve replacement. Ann Thorac Surg May 2008;85(5):1789−91.

[91] George KM, Smedira NG. Malpositioned mechanical mitral valves causing left ventricular outflow tract obstruction. J Heart Valve Dis March 2010;19(2):233−5.

[92] Blanke P, Naoum C, Dvir D, Bapat V, Ong K, Muller D, et al. Predicting LVOT obstruction in transcatheter mitral valve implantation: Concept of the neo-LVOT. JACC Cardiovasc Imaging April 2017;10(4):482−5.

[93] Paradis JM, Del Trigo M, Puri R, Rodes-Cabau J. Transcatheter valve-in-valve and valve-in-ring for treating aortic and mitral surgical prosthetic dysfunction. J Am Coll Cardiol November 3, 2015;66(18):2019−37.

[94] Cohen SR, Sell JE, McIntosh CL, Clark RE. Tricuspid regurgitation in patients with acquired, chronic, pure mitral regurgitation. II. Nonoperative management, tricuspid valve annuloplasty, and tricuspid valve replacement. J Thorac Cardiovasc Surg October 1987;94(4):488−97.

[95] Hahn RT. State-of-the-Art review of echocardiographic imaging in the evaluation and treatment of functional tricuspid regurgitation. Circ Cardiovasc Imaging December 2016;9(12):e005332.

[96] Rodes-Cabau J, Hahn RT, Latib A, Laule M, Lauten A, Maisano F, et al. Transcatheter therapies for treating tricuspid regurgitation. J Am Coll Cardiol April 19, 2016;67(15): 1829—45.

[97] Lee JW, Song JM, Park JP, Lee JW, Kang DH, Song JK. Long-term prognosis of isolated significant tricuspid regurgitation. Circ J February 2010;74(2):375—80.

[98] Benedetto U, Melina G, Angeloni E, Refice S, Roscitano A, Comito C, et al. Prophylactic tricuspid annuloplasty in patients with dilated tricuspid annulus undergoing mitral valve surgery. J Thorac Cardiovasc Surg March 2012;143(3):632—8.

[99] Kilic A, Saha-Chaudhuri P, Rankin JS, Conte JV. Trends and outcomes of tricuspid valve surgery in North America: an analysis of more than 50,000 patients from the Society of Thoracic Surgeons database. Ann Thorac Surg November 2013;96(5): 1546—52 [discussion 1552].

[100] Lindman BR, Maniar HS, Jaber WA, Lerakis S, Mack MJ, Suri RM, et al. Effect of tricuspid regurgitation and the right heart on survival after transcatheter aortic valve replacement: insights from the Placement of Aortic Transcatheter Valves II inoperable cohort. Circ Cardiovasc Interv April 2015;8(4). https://doi.org/10.1161/CIRCINTERVENTIONS.114.002073.

[101] Frangieh AH, Gruner C, Mikulicic F, Attinger-Toller A, Tanner FC, Taramasso M, et al. Impact of percutaneous mitral valve repair using the MitraClip system on tricuspid regurgitation. EuroIntervention April 8, 2016;11(14):e1680—6.

[102] Rogers JH, Bolling SF. The tricuspid valve: current perspective and evolving management of tricuspid regurgitation. Circulation May 26, 2009;119(20):2718—25.

[103] Fukuda S, Saracino G, Matsumura Y, Daimon M, Tran H, Greenberg NL, et al. Three-dimensional geometry of the tricuspid annulus in healthy subjects and in patients with functional tricuspid regurgitation: a real-time, 3-dimensional echocardiographic study. Circulation July 4, 2006;114(1 Suppl):I492—8.

[104] Muraru D, Veronesi F, Maddalozzo A, Dequal D, Frajhof L, Rabischoffsky A, et al. 3D printing of normal and pathologic tricuspid valves from transthoracic 3D echocardiography data sets. Eur Heart J Cardiovasc Imaging July 1, 2017;18(7):802—8.

[105] Amerini A, Hatam N, Malasa M, Pott D, Tewarie L, Isfort P, et al. A personalized approach to interventional treatment of tricuspid regurgitation: experiences from an acute animal study. Interact Cardiovasc Thorac Surg September 2014;19(3):414—8.

[106] Taramasso M, Phalla O, Spagnolo P, Guidotti A, Vicentini L, Scherman J, et al. 3D heart models for planning of percutaneous tricuspid interventions. J Am Coll Cardiol 2015;66(15):B53.

[107] O'Neill B, Wang DD, Pantelic M, Song T, Guerrero M, Greenbaum A, et al. Transcatheter caval valve implantation using multimodality imaging: roles of TEE, CT, and 3D printing. JACC Cardiovasc Imaging February 2015;8(2):221—5.

[108] Chung R, Taylor AM. Imaging for preintervention planning: transcatheter pulmonary valve therapy. Circ Cardiovasc Imaging January 2014;7(1):182—9.

[109] Pluchinotta FR, Bussadori C, Butera G, Piazza L, Chessa M, Saracino A, et al. Treatment of right ventricular outflow tract dysfunction: a multimodality approach. Eur Heart J Suppl April 28, 2016;18(Suppl E):E22—6.

[110] Schievano S, Migliavacca F, Coats L, Khambadkone S, Carminati M, Wilson N, et al. Percutaneous pulmonary valve implantation based on rapid prototyping of right ventricular outflow tract and pulmonary trunk from MR data. Radiology February 2007; 242(2):490—7.

[111] Armillotta A, Bonhoeffer P, Dubini G, Ferragina S, Migliavacca F, Sala G, et al. Use of rapid prototyping models in the planning of percutaneous pulmonary valved stent implantation. Proc Inst Mech Eng H May 2007;221(4):407—16.

[112] Phillips AB, Nevin P, Shah A, Olshove V, Garg R, Zahn EM. Development of a novel hybrid strategy for transcatheter pulmonary valve placement in patients following transannular patch repair of tetralogy of fallot. Catheter Cardiovasc Interv February 15, 2016;87(3):403—10.

[113] Valverde I, Sarnago F, Prieto R, Zunzunegui JL. Three-dimensional printing in vitro simulation of percutaneous pulmonary valve implantation in large right ventricular outflow tract. Eur Heart J April 21, 2017;38(16):1262—3.

Simulation of Percutaneous Structural Interventions

Serge C. Harb, Brian P. Griffin, L. Leonardo Rodriguez

Cleveland Clinic, Cleveland, Ohio, United States

INTRODUCTION

Recent applications of cardiovascular 3D printing have expanded to the use of the 3D printed models for simulation of percutaneous structural interventions, especially with the advent of flexible and transparent material. Procedural simulation on 3D printed models has been reported in a wide range of structural interventions including valvular (particularly aortic and mitral, but also pulmonic and tricuspid valves), congenital (both simple and complex congenital heart disease), vascular (especially for endovascular aortic procedures), and in miscellaneous other cardiac interventions (such as left atrial appendage closure [LAAC]). By providing physical models for interventional simulation, 3D printing has allowed improved procedural planning, device selection, risk identification, and training opportunities, all of which have been invaluable, particularly in cases of innovative and complex interventions.

Cardiovascular applications of 3D printing span multiple areas including:

- Diagnosis, via enhanced visualization and spatial appreciation of structures, especially in cases of complex congenital heart disease [1].
- Research [2], education [3], and patient communication [4].
- Simulation of interventions, both surgical and percutaneous structural [5]: a major advance in 3D printing has been the ability to print models in flexible and transparent material that allow the model to be bent, cut, and intervened upon with good visualization (given the transparency), providing the operator with the opportunity for patient-specific device bench testing.

Percutaneous structural interventions refer to noncoronary, nonsurgical cardiovascular interventions and have a wide range of applications that can be grouped into four major areas:

1. Valvular heart disease
2. Congenital heart disease
3. Vascular disease
4. Miscellaneous such as in LAAC, hypertrophic obstructive cardiomyopathy (HOCM), and cardiac tumors.

3D Printing Applications in Cardiovascular Medicine. https://doi.org/10.1016/B978-0-12-803917-5.00007-9

Table 7.1 Potential Benefits of Simulation Percutaneous Structural Interventions on 3D Printed Models

Reduction in procedural complications (for example, blood loss)
Reduction in procedural time (through enhanced planning such as device sizing on the 3D printed model)
Reduction in X-ray and contrast exposure
Possible reduction in procedural cost through reductions in procedural complications and time
Improved procedural efficiency
Improved patient communication
Training opportunities for junior physicians (residents and fellows)

Simulation of percutaneous structural interventions using 3D printed models has been applied to all four major areas, where it has enabled advanced procedural planning, device choice and sizing, and simulation of device delivery and deployment. Potential benefits of interventional simulation using 3D printed models are summarized in Table 7.1.

SIMULATION OF VALVULAR INTERVENTIONS

Percutaneous therapies for valvular heart disease have developed tremendously over the past decade [6]. In 2000 and 2002, the first catheter-based pulmonic and aortic valve replacement procedures were performed [7,8]. Since then, the number of percutaneous valve interventions has grown exponentially and continues to expand.

Three-dimensional printing in valvular heart disease has gained significant interest and is expected to emerge as an important technique to enhance the efficacy and safety of transcatheter valve therapy via improved procedural planning and interventional simulation. While more attention has been given to 3D printing of percutaneous therapies for left-sided valvular pathology, right-sided interventions and paravalvular leak closure can equally benefit from this technique.

AORTIC VALVE

Transcatheter aortic valve replacement (TAVR) has emerged as the mainstay of treatment for severe aortic stenosis in inoperable and high-risk patients, and recently its indication has expanded to intermediate risk patients [9].

Three-dimensional printing has had many applications in this area as a diagnostic, prognostic, training, and a simulation tool.

1. Diagnostic: Functional assessment of calcific aortic stenosis has been reproduced using 3D printed models [10,11].

2. Prognostic: Risk factors for postprocedural permanent pacemaker need were identified based on simulations of balloon valvuloplasty procedures performed on 3D printed annuli [12].
3. Training: Especially in cases of complex interventions, such as transapical TAVR [13].
4. Simulation: Valve in valve procedures are particularly challenging and simulation on 3D printed models can decrease the procedural risk and improve the prosthesis selection [14].

MITRAL VALVE

A number of technologies are currently in development for the percutaneous nonsurgical management of mitral regurgitation, including both percutaneous repair and replacement [15]. While only one technique has received FDA approval so far (the MitraClip), others are expected to make it to the market once safety and efficacy data become available.

Reported applications of 3D printed models in percutaneous mitral valve interventions include: guiding implantation of percutaneous repair systems [16], predicting the risk of left ventricular outflow tract obstruction after transcatheter mitral valve replacement [17], and simulation of mitral valve intervention on a 3D model to assist in selecting and sizing the occluder device [18].

TRICUSPID VALVE

Reports of tricuspid valve 3D printing for simulation of percutaneous interventions are more scarce compared to other valves, with one report of a 3D model of the vena cava-right atrium junction that was used for procedural planning of a caval valve implantation to treat functional tricuspid regurgitation [19]. However, with the development of percutaneous tricuspid valve therapies for tricuspid valve regurgitation [20], the role of 3D printing may become more prominent. Recently, at our institution, we have successfully performed the first implant of a transcatheter tricuspid valve stent (NaviGate®) in a native tricuspid valve with severe functional tricuspid regurgitation. The procedural planning relied heavily on 3D printing (Fig. 7.1), and repeated simulations on the 3D printed model were performed (Fig. 7.2). In another case of failed tricuspid surgical repair with an annuloplasty ring, a tricuspid valve stent was also successfully implanted percutaneously, via the superior vena cava, at the annular ring level. This procedure was also strongly dependent on 3D printing for procedural planning and interventional simulation (Fig. 7.3). The field of percutaneous tricuspid valve repair and replacement has a lot of potential and will continue to evolve over the years. Due to the more complex anatomy and geometry of the right heart, advanced imaging that will eventually translate into 3D printed constructs is expected to play a key role in the successful and efficacious delivery and deployment of such devices.

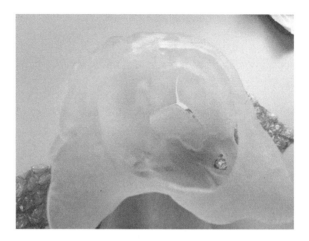

FIGURE 7.1

A 3D printed model of the right heart of a patient with severe functional tricuspid regurgitation. The model was used for the procedural planning of the transcatheter tricuspid valve implantation. The tricuspid valve with its three leaflets is seen in the top center of the model. The superior vena cava and inferior vena cava are seen from each side with the small hollow structure between the tricuspid valve and the inferior vena cava representing the coronary sinus.

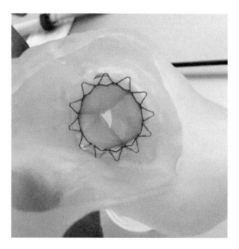

FIGURE 7.2

Simulation of the transcatheter tricuspid valve implantation on the 3D printed model. The prosthetic valve, as seen from an atrial perspective, deployed on the 3D printed model during one of the simulations of transcatheter tricuspid valve implantation performed on the model.

FIGURE 7.3

Simulation of percutaneous tricuspid valve implantation on the 3D printed model of a patient with severe tricuspid regurgitation after a failed tricuspid valve repair with an annuloplasty ring. The tricuspid annular ring was printed with a different, more rigid, white material as seen in the figure.

PULMONARY VALVE

Percutaneous pulmonary valve interventions are common in congenital pulmonary valve disease. Similar to other valves, intervention, simulation, and device selection based on 3D printed models have been reported [21] and the value of these models in patient selection for intervention has been demonstrated [22]. In fact, feasibility and success of percutaneous pulmonary valve interventions depend on the size of the right ventricle outflow tract in relation to the implant size, and 3D printed models of the outflow tract can provide detailed assessment of the implantation site and give insight on the most suitable prosthetic valve size.

PARAVALVULAR LEAKS

Paravalvular leaks are not uncommon after surgical or percutaneous valvular therapy, and when treatment is recommended, percutaneous closure approaches are attractive. Three-dimensional printed models not only help in predicting occurrence [23] of paravalvular leaks, but their percutaneous closure may also be simulated on physical models, helping determine the optimal interventional approach and appropriate closure device size [24].

SIMULATION OF CONGENITAL INTERVENTIONS

The heterogeneity and complexity of congenital heart disease makes 3D printing an ideal tool to help in the delivery of optimal personalized management strategies.

Percutaneous structural interventions play an important role in the management of both simple and complex congenital heart defects and 3D printing can be a helpful tool both for planning and simulation of these procedures.

SIMPLE CONGENITAL HEART DEFECTS

Atrial (ASD) and ventricular (VSD) septal defects are among the most common simple congenital heart defects. In cases where an intervention is recommended, both surgical and percutaneous approaches to close the defect are available. Multiple reports have emphasized the role of 3D printing, not only to better define the anatomy and understand the spatial relationships of neighboring structures, but also to plan the intervention and simulate the closure preoperatively [25,26]. In cases of ASD, the septal occluder used for percutaneous closure can be tested on a 3D printed model of the defect, or it can be printed after the procedure for quality management to assess its position and relation to surrounding structures [27].

Similarly, in cases of VSD, both congenital and acquired (for example, post myocardial infarction), surgical patching or percutaneous closure can be practiced on a 3D printed model of the defect, allowing simulation of the intervention and more accurate sizing of the closure device [28].

COMPLEX CONGENITAL HEART DEFECTS

While 3D printing has been used extensively as an adjunctive diagnostic tool in cases of complex congenital diseases [29,30], it can also play an important role in simulation of the interventional procedures, both surgical and percutaneous. These interventions are typically challenging, given the anatomical complexity of the defects, and the opportunity to practice on the benchtop provided by 3D printing can be a major determinant of procedural success. Interventional simulation on 3D printed models has already been shown to be useful in some challenging cases of transposition of the great vessels [31], hypoplastic heart syndrome [32], and stent angioplasty of pulmonary venous baffle obstruction [33].

SIMULATION OF VASCULAR INTERVENTIONS

Three-dimensional printing of various vascular beds has already been reported including the aortic root and ascending aorta with its great vessels [34] and the abdominal aorta with its branch vessels [35]. In challenging cases, preprocedural planning of a surgical intervention and/or vascular device deployment relying on a 3D printed model of the patient's specific vasculature can be beneficial.

This has been demonstrated in complex aortic procedures such as endovascular stent interventions on aortic arch hypoplasia [36] and in the frozen elephant trunk surgery [37]. Three-dimensional printing has also allowed the development of an alternative to aortic root replacement in patients with Marfan syndrome [38,39], where an aortic 3D printed model spanning from the root to the proximal arch is used to provide a patient's tailored support device.

The added value of 3D printing has also been demonstrated in endovascular abdominal aortic aneurysm repair in various beds: celiac [40], splenic [41], and renal artery aneurysms [35].

Typically, the vascular model is printed hollow and with flexible material, in order to allow preprocedural simulation.

SIMULATION OF MISCELLANEOUS PERCUTANEOUS INTERVENTIONS

LEFT ATRIAL APPENDAGE CLOSURE

Percutaneous LAAC has become the treatment of choice for stroke prevention in patients with nonvalvular atrial fibrillation, at high risk of bleeding [42]. Multiple devices are available, with one FDA approved (Watchman). While each of the devices comes in various sizes, the complex anatomy of the appendage and its variable morphology may result in suboptimal appendage occlusion [43] with resultant failure in realizing its stroke preventive role. Also, inaccurate sizing leads to an increased risk of complications [44]. Various reports [45,46] have shown that three-dimensional printing is able to provide a more precise procedural planning. The 3D printed model of the patient's specific left atrial appendage allows selection of the optimal device size. In addition, adding both atria and interatrial septum to the 3D printed model allows simulation of the delivery process with selection of the optimal site of transseptal puncture that would allow easy deliver of the device. Fig. 7.4 shows a 3D model of a left atrial appendage with both atria and interatrial septum that we used, at our institution, for procedural simulation prior to Watchman implantation.

Adding to the advantages cited above is the opportunity for physicians in training to practice device implantation. This has already been shown to improve the learning curve of early operators [47].

HYPERTROPHIC OBSTRUCTIVE CARDIOMYOPATHY

HOCM is a genetic heart muscle disease characterized by asymmetric left ventricular hypertrophy. Patients with outflow tract obstruction secondary to septal hypertrophy can remain symptomatic despite optimal medical therapy. Management strategies at that stage include septal myectomy and alcohol septal ablation. These

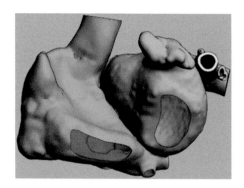

FIGURE 7.4

Three-dimensional model of a left atrial appendage with both atria and interatrial septum used for procedural simulation prior to Watchman implantation. Ventricular view of the upper heart chambers. The right atrium is in blue and the left atrium is in red with the interatrial septum in between. The ventricles are not represented. The left atrial appendage is clearly seen above the left-sided pulmonary veins.

interventions are complex and outcomes highly dependent on the center's and operator's volume and experience [48]. Three-dimensional printing has been used in surgical planning [49], allowing enhanced understanding of the left ventricular anatomy, and also for preoperational simulation of the myectomy [50]. The model was printed using flexible, rubberlike, and transparent materials allowing the surgeon to handle and disassemble the myocardial 3D model.

OTHER MISCELLANEOUS APPLICATIONS

Three-dimensional models of cardiac aneurysms [24] and cardiac tumors [51,52] have been used to better understand the anatomy and assist in the choice of the optimal therapeutic approach. Simulation of the procedure on a 3D printed model would help in the proper selection of the therapeutic intervention, along with the material to use (catheter shape in case of a percutaneous approach) and delivery strategy (optimal route of access).

CONCLUSION

In conclusion, 3D printing has the potential to revolutionize the field of percutaneous structural interventions. The physical models allow procedural simulation and bench practicing, which could be major determinants for success, particularly in complex innovative interventions.

REFERENCES

[1] Mitsouras D, Liacouras P, Imanzadeh A, et al. Medical 3D printing for the radiologist. Radiographics November–December 2015;35(7):1965–88.

[2] Mashari A, Knio Z, Jeganathan J, et al. Hemodynamic testing of patient-specific mitral valves using a pulse duplicator: a clinical application of three-dimensional printing. J Cardiothorac Vasc Anesth October 2016;30(5):1278–85.

[3] Costello JP, Olivieri LJ, Su L, et al. Incorporating three-dimensional printing into a simulation-based congenital heart disease and critical care training curriculum for resident physicians. Congenit Heart Dis March–April 2015;10(2):185–90.

[4] Biglino G, Capelli C, Wray J, et al. 3D-manufactured patient-specific models of congenital heart defects for communication in clinical practice: feasibility and acceptability. BMJ Open April 30, 2015;5(4):e007165.

[5] Schmauss D, Haeberle S, Hagl C, Sodian R. Three-dimensional printing in cardiac surgery and interventional cardiology: a single-centre experience. Eur J Cardio Thorac Surg June 2015;47(6):1044–52.

[6] Feldman T, Young A. Percutaneous approaches to valve repair for mitral regurgitation. J Am Coll Cardiol May 27, 2014;63(20):2057–68.

[7] Bonhoeffer P, Boudjemline Y, Saliba Z, et al. Percutaneous replacement of pulmonary valve in a right-ventricle to pulmonary-artery prosthetic conduit with valve dysfunction. Lancet October 21, 2000;356(9239):1403–5.

[8] Cribier A, Eltchaninoff H, Bash A, et al. Percutaneous transcatheter implantation of an aortic valve prosthesis for calcific aortic stenosis: first human case description. Circulation December 10, 2002;106(24):3006–8.

[9] Nishimura RA, Otto CM, Bonow RO, et al. 2014. AHA/ACC guideline for the management of patients with valvular heart disease: A report of the american college of cardiology/american heart association task force on practice guidelines. J Am Coll Cardiol 2014;63:e57–185.

[10] Maragiannis D, Jackson MS, Igo SR, Chang SM, Zoghbi WA, Little SH. Functional 3D printed patient-specific modeling of severe aortic stenosis. J Am Coll Cardiol September 09, 2014;64(10):1066–8.

[11] Maragiannis D, Jackson MS, Igo SR, et al. Replicating patient-specific severe aortic valve stenosis with functional 3D modeling. Circ Cardiovasc Imaging October 2015; 8(10):e003626.

[12] Fujita B, Kutting M, Seiffert M, et al. Calcium distribution patterns of the aortic valve as a risk factor for the need of permanent pacemaker implantation after transcatheter aortic valve implantation. Eur Heart J Cardiovasc Imaging December 2016;17(12):1385–93.

[13] Abdel-Sayed P, Kalejs M, von Segesser LK. A new training set-up for trans-apical aortic valve replacement. Interact Cardiovasc Thorac Surg June 2009;8(6):599–601.

[14] Fujita B, Kutting M, Scholtz S, et al. Development of an algorithm to plan and simulate a new interventional procedure. Interact Cardiovasc Thorac Surg July 2015;21(1): 87–95.

[15] Figulla HR, Webb JG, Lauten A, Feldman T. The transcatheter valve technology pipeline for treatment of adult valvular heart disease. Eur Heart J July 21, 2016;37(28): 2226–39.

[16] Dankowski R, Baszko A, Sutherland M, et al. 3D heart model printing for preparation of percutaneous structural interventions: description of the technology and case report. Kardiol Pol 2014;72(6):546—51.

[17] Wang DD, Eng M, Greenbaum A, et al. Predicting LVOT obstruction after TMVR. JACC Cardiovasc Imaging November 2016;9(11):1349—52.

[18] Little SH, Vukicevic M, Avenatti E, Ramchandani M, Barker CM. 3D printed modeling for patient-specific mitral valve intervention: repair with a clip and a plug. JACC Cardiovasc Interv May 09, 2016;9(9):973—5.

[19] O'Neill B, Wang DD, Pantelic M, et al. Transcatheter caval valve implantation using multimodality imaging: roles of TEE, CT, and 3D printing. JACC Cardiovasc Imaging February 2015;8(2):221—5.

[20] Taramasso M, Pozzoli A, Guidotti A, et al. Percutaneous tricuspid valve therapies: the new frontier. Eur Heart J March 01, 2017;38(9):639—47.

[21] Chung R, Taylor AM. Imaging for preintervention planning: transcatheter pulmonary valve therapy. Circ Cardiovasc Imaging January 2014;7(1):182—9.

[22] Schievano S, Migliavacca F, Coats L, et al. Percutaneous pulmonary valve implantation based on rapid prototyping of right ventricular outflow tract and pulmonary trunk from MR data. Radiology February 2007;242(2):490—7.

[23] Ripley B, Kelil T, Cheezum MK, et al. 3D printing based on cardiac CT assists anatomic visualization prior to transcatheter aortic valve replacement. J Cardiovasc Comput Tomogr January—February 2016;10(1):28—36.

[24] Kim MS, Hansgen AR, Wink O, Quaife RA, Carroll JD. Rapid prototyping: a new tool in understanding and treating structural heart disease. Circulation May 06, 2008; 117(18):2388—94.

[25] Shiraishi I, Yamagishi M, Hamaoka K, Fukuzawa M, Yagihara T. Simulative operation on congenital heart disease using rubber-like urethane stereolithographic biomodels based on 3D datasets of multislice computed tomography. Eur J Cardio Thorac Surg February 2010;37(2):302—6.

[26] Samuel BP, Pinto C, Pietila T, Vettukattil JJ. Ultrasound-derived three-dimensional printing in congenital heart disease. J Digit Imaging August 2015;28(4):459—61.

[27] Bartel T, Rivard A, Jimenez A, Edris A. Three-dimensional printing for quality management in device closure of interatrial communications. Eur Heart J Cardiovasc Imaging September 2016;17(9):1069.

[28] Lazkani M, Bashir F, Brady K, Pophal S, Morris M, Pershad A. Postinfarct VSD management using 3D computer printing assisted percutaneous closure. Indian Heart J November—December 2015;67(6):581—5.

[29] Ryan JR, Moe TG, Richardson R, Frakes DH, Nigro JJ, Pophal S. A novel approach to neonatal management of tetralogy of Fallot, with pulmonary atresia, and multiple aortopulmonary collaterals. JACC Cardiovasc Imaging January 2015;8(1):103—4.

[30] Riesenkampff E, Rietdorf U, Wolf I, et al. The practical clinical value of three-dimensional models of complex congenitally malformed hearts. J Thorac Cardiov Sur September 2009;138(3):571—80.

[31] Valverde I, Gomez G, Gonzalez A, et al. Three-dimensional patient-specific cardiac model for surgical planning in Nikaidoh procedure. Cardiol Young April 2015;25(4): 698—704.

[32] Shiraishi I, Kajiyama Y, Yamagishi M, Hamaoka K. Images in cardiovascular medicine. Stereolithographic biomodeling of congenital heart disease by multislice computed tomography imaging. Circulation May 02, 2006;113(17):e733—734.

[33] Olivieri L, Krieger A, Chen MY, Kim P, Kanter JP. 3D heart model guides complex stent angioplasty of pulmonary venous baffle obstruction in a mustard repair of D-TGA. Int J Cardiol March 15, 2014;172(2):e297—298.

[34] Pepper J, Petrou M, Rega F, Rosendahl U, Golesworthy T, Treasure T. Implantation of an individually computer-designed and manufactured external support for the Marfan aortic root. Multimed Man Cardiothorac Surg 2013;2013:mmt004.

[35] Giannopoulos AA, Steigner ML, George E, et al. Cardiothoracic applications of 3-dimensional printing. J Thorac Imaging September 2016;31(5):253—72.

[36] Valverde I, Gomez G, Coserria JF, et al. 3D printed models for planning endovascular stenting in transverse aortic arch hypoplasia. Catheter Cardiovasc Interv May 2015; 85(6):1006—12.

[37] Schmauss D, Juchem G, Weber S, Gerber N, Hagl C, Sodian R. Three-dimensional printing for perioperative planning of complex aortic arch surgery. Ann Thorac Surg June 2014;97(6):2160—3.

[38] Golesworthy T, Lamperth M, Mohiaddin R, Pepper J, Thornton W, Treasure T. The Tailor of Gloucester: a jacket for the Marfan's aorta. Lancet October 30—November 5 2004;364(9445):1582.

[39] Pepper J, Golesworthy T, Utley M, et al. Manufacturing and placing a bespoke support for the Marfan aortic root: description of the method and technical results and status at one year for the first ten patients. Interact Cardiovasc Thorac Surg March 2010;10(3): 360—5.

[40] Salloum C, Lim C, Fuentes L, Osseis M, Luciani A, Azoulay D. Fusion of information from 3D printing and surgical robot: an innovative minimally technique illustrated by the resection of a large celiac trunk aneurysm. World J Surg January 2016;40(1):245—7.

[41] Itagaki MW. Using 3D printed models for planning and guidance during endovascular intervention: a technical advance. Diagn Interv Radiol July—August 2015;21(4):338—41.

[42] Holmes Jr DR, Lakkireddy DR, Whitlock RP, Waksman R, Mack MJ. Left atrial appendage occlusion: opportunities and challenges. J Am Coll Cardiol February 04, 2014;63(4):291—8.

[43] Unsworth B, Sutaria N, Davies DW, Kanagaratnam P. Successful placement of left atrial appendage closure device is heavily dependent on 3-dimensional transesophageal imaging. J Am Coll Cardiol September 13, 2011;58(12):1283.

[44] Pison L, Potpara TS, Chen J, et al. Left atrial appendage closure-indications, techniques, and outcomes: results of the European Heart Rhythm Association Survey. Europace April 2015;17(4):642—6.

[45] Pellegrino PL, Fassini G, Di Biase M, Tondo C. Left atrial appendage closure guided by 3D printed cardiac reconstruction: emerging directions and future trends. J Cardiovasc Electrophysiol June 2016;27(6):768—71.

[46] Otton JM, Spina R, Sulas R, et al. Left atrial appendage closure guided by personalized 3d-printed cardiac reconstruction. JACC Cardiovasc Interv June 2015;8(7):1004—6.

[47] Wang DD, Eng M, Kupsky D, et al. Application of 3-dimensional computed tomographic image guidance to WATCHMAN implantation and impact on early operator learning curve: single-center experience. JACC Cardiovasc Interv November 28, 2016;9(22):2329—40.

[48] Kim LK, Swaminathan RV, Looser P, et al. Hospital volume outcomes after septal myectomy and alcohol septal ablation for treatment of obstructive hypertrophic cardiomyopathy: US nationwide inpatient database, 2003—2011. JAMA Cardiol June 01, 2016;1(3):324—32.

[49] Hermsen JL, Burke TM, Seslar SP, et al. Scan, plan, print, practice, perform: development and use of a patient-specific 3-dimensional printed model in adult cardiac surgery. J Thorac Cardiovasc Surg January 2017;153(1):132−40.

[50] Yang DH, Kang JW, Kim N, Song JK, Lee JW, Lim TH. Myocardial 3-dimensional printing for septal myectomy guidance in a patient with obstructive hypertrophic cardiomyopathy. Circulation July 28, 2015;132(4):300−1.

[51] Schmauss D, Gerber N, Sodian R. Three-dimensional printing of models for surgical planning in patients with primary cardiac tumors. J Thorac Cardiovasc Surg May 2013;145(5):1407−8.

[52] Al Jabbari O, Abu Saleh WK, Patel AP, Igo SR, Reardon MJ. Use of three-dimensional models to assist in the resection of malignant cardiac tumors. J Card Surg September 2016;31(9):581−3.

4D Printing of Actuating Cardiac Tissue

Vahid Serpooshan[1,2,3]**, James B. Hu**[1]**, Orlando Chirikian**[1]**, Daniel A. Hu**[1]**,
Morteza Mahmoudi**[4]**, Sean M. Wu**[1,5]

*Stanford Cardiovascular Institute, Stanford University School of Medicine, Stanford, California,
United States*[1]*; Department of Biomedical Engineering, Georgia Institute of Technology, Atlanta,
Georgia, United States*[2]*; Emory University School of Medicine, Atlanta, Georgia, United States*[3]*;
Department of Anesthesiology, Brigham & Women's Hospital, Harvard Medical School, Boston,
Massachusetts, United States*[4]*; Department of Medicine, Division of Cardiovascular Medicine,
Stanford University, Stanford, California, United States*[5]

INTRODUCTION

Tissue engineering is at the intersection of a multitude of disciplines including, but not limited to, bioengineering, cell biology, mechanical engineering, material science, and computer science. Recent advances in adjunct fields have allowed the tissue engineering discipline to see unprecedented growth, allowing researchers to actualize their collective vision of creating biomimetic tissue constructs for medical and pharmaceutical applications. Traditional three-dimensional (3D) bioprinting has made great strides in its attempt to create such tissue constructs [1−4], but falls short because of its static nature [5,6]. Native tissues are in a state of flux, constantly changing and adapting in response to their environment. Thus, four-dimensional (4D) printing or shape-morphing systems are being developed to address this salient concern (Fig. 8.1).

There are many working definitions of 4D printing with subtle differences between them, but what unifies all these definitions is the additional time dimension. 4D printing refers to the ability of fabricated constructs to adapt in response to external and/or internal stimuli, allowing for dynamic changes in shape, physical properties, or functionality to adjust to small changes in their environment [7−10]. Although a nascent field of research, there have been multiple successful attempts in printing stimuli-responsive constructs. One approach is to utilize smart materials that are sensitive to stimuli such as temperature, electricity, cell-traction forces, humidity, and pH [11−15]. There is a plethora of smart materials that can respond to different stimuli and be adapted for 4D printing. Here, we will review temperature-sensitive, electrically-sensitive, and cell-traction force-sensitive smart materials; discuss their use in creating 4D structures; and briefly explore some current applications in creating actuating cardiac tissue.

3D Printing Applications in Cardiovascular Medicine. https://doi.org/10.1016/B978-0-12-803917-5.00008-0

153

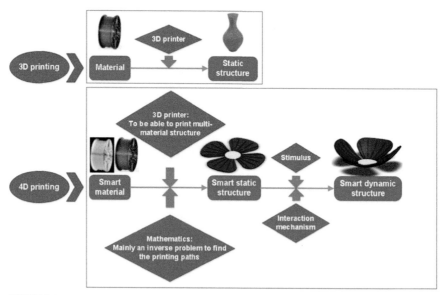

FIGURE 8.1

The key differences between 3D and 4D printing. 3D printing involves deposition of material into a predetermined static shape. 4D printing, on the other hand, involves careful deposition of a smart material into a predetermined, smart static structure. When this smart static structure interacts with an internal or external stimulus, it will transform its shape and become a smart dynamic structure.

Reproduced with permission from Momeni, F. et al. A review of 4D printing. Mater Des 2017;122:42–79.

TEMPERATURE-SENSITIVE MATERIALS

Thermo-responsive polymers are one of the most widely employed smart materials. This class of materials has the characteristic ability to change shape or volume in response to temperature changes [16]. Often, these materials exhibit shape memory, the ability for the material to recollect its original shape when the external temperature stimulus is removed [17,18]. A canonical example of such thermo-responsive polymers is Poly(N-isopropylacrylamide) (pNIPAM). When the temperature is below the critical solution temperature, the pNIPAM polymer chains become hydrophilic; when heated to higher temperatures (around 32°C depending on how the polymer is functionalized), the polymer undergoes a lower critical solution temperature phase transition. The polymer chains become hydrophobic and the polymer dehydrates [19–21].

Many research groups have employed this polymer or its derivatives to achieve shape transformations. Okuzaki and colleagues developed a nanofiber material that incorporated pNIPAM. Their study demonstrated that temperature changes could modulate shrinking of nanofiber patch [22]. Additionally, when oscillating between two specific temperatures, the material would cycle between shrunken and normal volume states. The fabrication of more complex structures has been explored with this particular polymer. Breger et al. created a derivative of pNIPAM by functionalizing it with polypropylene fumarate [23]. Using this temperature-sensitive polymer, investigators achieved more elaborate shape transformation. When heated to 36°C, the floral-shaped polymer construct would close in on itself and form a functional claw and grip other objects (Fig. 8.2). Similar to how the primary structure of amino acids determine the ultimate tertiary and quaternary structure of protein, the initial geometries of the temperature-sensitive construct will determine how it will fold. Since the mechanism of action by which such constructs fold is well understood, predictive models may be integrated in the future to rapidly create the intricate final designs of these constructs. pNIPAM by itself cannot be used for printing; instead it can be combined with other materials to achieve proper rheological properties for printing. Bakarich and colleagues used an alginate and pNIPAM mixture to achieve a printable hydrogel that can thermally actuate, exhibiting swelling and shrinking cycles [24]. Besides shape alteration, other transformations

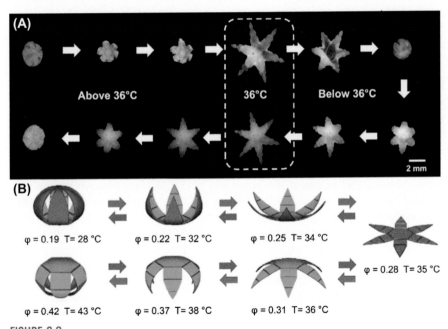

FIGURE 8.2

Experimental and theoretical images of a smart thermosensitive construct (microgripper) at varying temperatures. (A) As the temperature is dropped below 36°C, the thermosensitive microgripper closes, but this self-folding can be reversed upon heating. (B) Theoretical (simulation) images, corresponding to the experimentally acquired images, indicate that this folding action depends on temperature and swelling function.

Reproduced with permission from Breger, JC et al. Self-folding thermo-magnetically responsive soft microgrippers. ACS Appl Mater Interfaces 2015;7:3398–405.

can be made. For example, Li et al. developed a printable hydrogel composite of pNIPAM, poly(acrylamide), and carbon nanodots that can change its photoluminescence in response to temperature, a property that can be used to develop smart temperature sensors [25].

Additionally, temperature-sensitive polymers are of interest because they often function at the temperature range of the human body [26]. However, these polymers must be combined with other copolymers to be more biocompatible to cells. By fine-tuning the physical properties of these polymers and adding extra layers of functionality, these smart materials can be synchronized with the temperature fluxes inherent in biological systems to improve cellular performance. Hsieh and colleagues developed a biodegradable polyurethane temperature-sensitive hydrogel that was used to encapsulate and print murine neural stem cells and assist proliferation and differentiation in 3D [27]. Similarly, Kesti et al. utilized a pNIPAM and methacrylated hyaluronan polymer to create a fast-gelling bioink that encapsulated bovine chondrocytes. This polymer combination provided strong structural fidelity post print and supported cell viability for at least 1 week [28]. Currently, there are no 4D printing of actuating cardiac tissue utilizing thermo-responsive materials, but we envision that hybrid polymer blends can be created that are printable, biocompatible, and exhibit time-dependent transformations leading to improved tissue functionality.

ELECTRICAL-SENSITIVE (CONDUCTIVE) MATERIALS

Most electrical-responsive smart materials are characteristically conductive. Two widely used materials are polyalanine and polypyrrole, both of which are biocompatible [29—31]. To achieve electrical conductivity, a dopant/strong acid is used to protonate the backbone of the polyalanine [32,33]. Once doped, the polymer can be used to make films combined with other biocompatible materials to modulate the mechanical and chemical properties and tune it to that of native tissues. For instance, when combined with polycaprolactone (PCL), polyalanine can be electrospun to electrically active fibers that can support cardiomyocyte viability and function (Fig. 8.3) [34]. Polypyrrole can be functionalized by doping the polymer with biotin. Once added, these conductive surfaces can be electrically stimulated to release inert gold nanoparticles [35]. This system can be employed for controlled drug release, and may be adapted to help support cardiac tissue constructs as well.

Electrically sensitive materials are particularly relevant in cardiovascular tissue engineering, since the mammalian heart may be subjected to electrical control by pacing. Electrical signals begin at the sinoatrial node and propagate throughout the rest of the heart, passing through the atrioventricular node and Purkinje fibers; these pulses are rhythmic and synchronized. Many researchers have employed conductive polymers in construction of cardiac tissue scaffolds in order to create tissue scaffolds that are more biomimetic and functional. Hsiao and colleagues fabricated nanofiber meshes with PLGA or poly(lactic-co-glycolic acid) and polyalanine and reported that cardiomyocytes seeded onto the mesh showed enhanced coupling, synchronized beating, and adhered with higher affinity due to electrical interactions

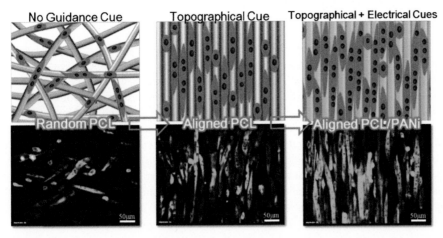

FIGURE 8.3

Electrospun nanofibers of polyaniline and poly(ε-caprolactone) provide electrical and topographical cues that aid in alignment of mouse skeletal myoblasts. Aligned nonconductive nanofibers guide myoblast orientation and promote myotube formation, while electrically conductive nanofibers further increase myotube maturation.

Reproduced with permission from Chen, MC et al. Electrically conductive nanofibers with highly oriented structures and their potential application in skeletal muscle tissue engineering. Acta Biomater 2013;9:5562–72.

between the conductive mesh and proteins on the cell surface. Additionally, these constructs can be field-stimulated by applying an external electrical current to control the contraction rate [36]. The application of field-stimulation to engineered cardiac tissue has been widely used to increase cell alignment and coupling [37].

Other conductive biocompatible materials have also been explored for bioprinting applications. Mixtures of hydrogels with conductive carbon nanotubes have been used to create structures that support cell growth and adhesion, while being electrically conductive [38]. Durmus et al. utilized a piezoelectric droplet printer to develop a 3D patterned network of lipids functionalized to allow for specific electrical propagation [39]. Additionally, these structures can use osmolality gradients in the hydrogel to promote specific folding of the network. If programmed correctly, these materials show great promise in making the future of electrically responsive 4D bioprinting a reality.

CELL-TRACTION FORCE-SENSITIVE MATERIALS

Rather than using external stimuli, such as temperature or electrical stimulation, another method to induce shape transformation is to use the contractile forces available to cells due to actin's interactions with myosin [40]. This is the basis for the development of the cell origami technique that allows researchers to create structures that can self-assemble to form 3D structures. By seeding cells onto two-dimensional plates in various configurations and concentrations, Takeuchi et al. have been able to fabricate complex geometric structures like cubes, dodecahedrons, and helical tubes [13]. Many of these interesting geometries can be applied to cardiovascular tissue engineering. For example, the helical tube geometry may be conducive to the engineering of blood vessels. Integrated vasculature is necessary for the fabrication of any large tissue constructs [41—43]. Cells in these structures cannot maintain their basic metabolic functions without having vasculature available to provide nutrients and oxygen, and shuttle waste away [44]. Vascular structures are difficult to engineer because of their small varying diameters, which undermine the fidelity of the construct. Thus, traditional methods to create these structures with 3D bioprinting employ either a support bath or a sacrificial material to hold the structure in place [45]. High-throughput fabrication using these traditional techniques is challenging and labor-intensive. On the other hand, cell origami may serve as a viable alternative. By using endothelial cells for the cell-origami process, tubular structures can be quickly self-assembled, which can later be used for the creation of vasculature. This is but one example of the application of this technology. Cell origami can be broadly applied to 4D printing technology to create structures with complex architecture (Fig. 8.4).

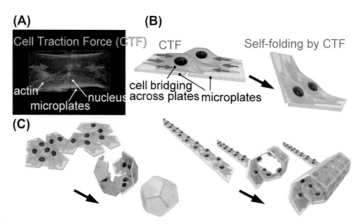

FIGURE 8.4

Conceptual illustration of cell origami. (A) When cells adhere across two microplates, the overall cell traction force brings two plates towards one another. Green and blue colors show actin and nucleus, respectively. (B) Parylene microplates are seeded with cells and fold in response to traction forces provided by the cells. (C) A wide variety of microstructures can be created by rearranging the configuration of the plates.

Reproduced with permission from Kuribayashi-Shigetomi, K et al. Cell origami: self-folding of three-dimensional cell-laden microstructures driven by cell traction force. PLoS One 2012;7:e51085.

BIOENGINEERING OF ACTUATING CARDIAC TISSUES

Feinberg et al. have explored creating cardiac tissues with functional materials in a 2D setting using thin muscular films. Using a thermally sensitive hydrogel to achieve 3D conformations, they were able to perform a variety of tasks, such as gripping and walking [46]. Besides performing mechanical tasks, these engineered cardiac micro-tissue constructs serve as a means to study the mechanics of cardiac tissue and as a possible drug screening platform [47]. When combined with 3D bioprinting, they can create functional 3D constructs with a variety of smart bioinks in a single programmable printing process [2,48].

The concept of 4D printing for cardiac tissue engineering is in its infancy. Challenges in our ability to synthesize smart hydrogels that are printable, biocompatible, and functional have limited the pace of progress in this field. Lind et al. recently made significant strides in developing a cardiac tissue microsystem that utilizes a variety of functional inks that help guide the assembly of cardiac cells and integrated sensors that give quantitative electrical readouts of cardiac cell contractions [49]. The authors also conducted a series of drug studies as a proof of concept that such a platform could be used for high-throughput drug screening. These early developments in 4D cardiac printing have focused less on creating clinically useful constructs, but rather on developing platforms to study cardiac tissue for drug screening and disease modeling.

Interestingly, Park et al. recently explored the use of genetically modified cardiomyocytes to generate a soft-tissue robot. By employing optogenetics technology in engineered cardiomyocytes, these investigators created a sting-ray-shaped device that can propel itself forward in water in response to optogenetic stimulation. The robot itself was made up of multiple layers of cardiomyocytes, each contributing to the overall forward propulsion of the sting-ray [50]. This level of intricacy is needed to construct clinically relevant cardiac tissue.

CONCLUSION

In summary, the rapidly growing range of printable materials, improved software and hardware design, and decreasing cost have resulted in the development of a new additive manufacturing technology called 4D printing, where printed objects have the ability to change form or function with time as a response to various stimuli such as heat, water, current, or light. 4D printing may be of great significance in the future due to its potential to redefine design and manufacturing principles. To replace conventional manufacturing methods, however, this technology still requires further refinement and optimization in areas including printing modality, scalability, materials (particularly biocompatible, responsive materials), and product market.

REFERENCES

[1] Murphy SV, Atala A. 3D bioprinting of tissues and organs. Nat Biotechnol 2014;32: 773–85. https://doi.org/10.1038/nbt.2958.

[2] Kolesky DB, Truby RL, Gladman AS, Busbee TA, Homan KA, Lewis JA. 3D bioprinting of vascularized, heterogeneous cell-laden tissue constructs. Adv Mater 2014; 26:3124–30. https://doi.org/10.1002/adma.201305506.

[3] Kang H-W, Lee SJ, Ko IK, Kengla C, Yoo JJ, Atala AA. 3D bioprinting system to produce human-scale tissue constructs with structural integrity. Nat Biotechnol 2016;34: 312–9. https://doi.org/10.1038/nbt.3413.

[4] Serpooshan V, Mahmoudi M, Hu DA, Hu JB, Wu SM. Bioengineering cardiac constructs using 3D printing. J 3D Print Med 2017;1:123–39. https://doi.org/10.2217/3dp-2016-0009.

[5] Gao B, Yang Q, Zhao X, Jin G, Ma Y, Xu F. 4D bioprinting for biomedical applications. Trends Biotechnol 2016;34:746–56. https://doi.org/10.1016/j.tibtech.2016.03.004.

[6] Mironov V, Reis N, Derby B. Review: bioprinting: a beginning. Tissue Eng 2006;12: 631–4. https://doi.org/10.1089/ten.2006.12.631.

[7] Tibbits S. 4D printing: multi-material shape change. Archit Design 2014;84:116–21. https://doi.org/10.1002/ad.1710.

[8] Ge Q, Qi HJ, Dunn ML. Active materials by four-dimension printing. Appl Phys Lett 2013;103:131901. https://doi.org/10.1063/1.4819837.

[9] Khoo ZX, Teoh JEM, Liu Y, Chua CK, Yang S, An J, et al. 3D printing of smart materials: a review on recent progresses in 4D printing. Virtual Phys Prototyp 2015;10: 103–22. https://doi.org/10.1080/17452759.2015.1097054.

[10] Sydney Gladman A, Matsumoto EA, Nuzzo RG, Mahadevan L, Lewis JA. Biomimetic 4D printing. Nat Mater 2016;15:413–8. https://doi.org/10.1038/nmat4544.

[11] Kim Y-J, Matsunaga YT. Thermo-responsive polymers and their application as smart biomaterials. J Mater Chem B 2017;5:4307–21. https://doi.org/10.1039/C7TB00157F.

[12] Balint R, Cassidy NJ, Cartmell SH. Conductive polymers: towards a smart biomaterial for tissue engineering. Acta Biomater 2014;10:2341–53. https://doi.org/10.1016/j.actbio.2014.02.015.

[13] Kuribayashi-Shigetomi K, Onoe H, Takeuchi S. Cell origami: self-folding of three-dimensional cell-laden microstructures driven by cell traction force. PLoS One 2012; 7:e51085. https://doi.org/10.1371/journal.pone.0051085.

[14] Kim HC, Mun S, Ko H-U, Zhai L, Kafy A, Kim J. Renewable smart materials. Smart Mater Struct 2016;25:73001. https://doi.org/10.1088/0964-1726/25/7/073001.

[15] Zhang Z, Chen L, Zhao C, Bai Y, Deng M, Shan H, et al. Thermo- and pH-responsive HPC-g-AA/AA hydrogels for controlled drug delivery applications. Polymer 2011;52: 676–82. https://doi.org/10.1016/j.polymer.2010.12.048.

[16] Ayano E, Kanazawa H. Temperature-responsive smart packing materials utilizing multi-functional polymers. Anal Sci 2014;30:167–73. https://doi.org/10.2116/analsci.30.167.

[17] Behl M, Lendlein A. Shape-memory polymers. Mater Today 2007;10:20–8. https://doi.org/10.1016/S1369-7021(07)70047-0.

[18] Wadood A. Brief overview on nitinol as biomaterial. Adv Mater Sci Eng 2016. https://www.hindawi.com/journals/amse/2016/4173138/.

[19] Soppimath KS, Aminabhavi TM, Dave AM, Kumbar SG, Rudzinski WE. Stimulus-responsive "smart" hydrogels as novel drug delivery systems. Drug Dev Ind Pharm 2002;28:957–74. https://doi.org/10.1081/DDC-120006428.

[20] Ochi M, Ida J, Matsuyama T, Yamamoto H. Effect of synthesis temperature on characteristics of PNIPAM/alginate IPN hydrogel beads. J Appl Polym Sci 2015;132. https://doi.org/10.1002/app.41814. n/a.

[21] Coronado R, Pekerar S, Lorenzo AT, Sabino MA. Characterization of thermo-sensitive hydrogels based on poly(N-isopropylacrylamide)/hyaluronic acid. Polym Bull 2011;67:101–24. https://doi.org/10.1007/s00289-010-0407-6.

[22] Okuzaki H, Kobayashi K, Hishiki F, Su S-J, Yan H. Thermo-responsive nanofiber mats fabricated by electrospinning of poly(N-isopropylacrylamide-co-stearyl acrylate). J Nanosci Nanotechnol 2011;11:5193–8.

[23] Breger JC, Yoon C, Xiao R, Kwag HR, Wang MO, Fisher JP, et al. Self-folding thermo-magnetically responsive soft microgrippers. ACS Appl Mater Interfaces 2015;7:3398–405. https://doi.org/10.1021/am508621s.

[24] Bakarich SE, Gorkin R, Panhuis M in het, Spinks GM. 4D printing with mechanically robust, thermally actuating hydrogels. Macromol Rapid Commun 2015;36:1211–7. https://doi.org/10.1002/marc.201500079.

[25] Li P, Huang L, Lin Y, Shen L, Chen Q, Shi W. Printable temperature-responsive hybrid hydrogels with photoluminescent carbon nanodots. Nanotechnology 2014;25:55603. https://doi.org/10.1088/0957-4484/25/5/055603.

[26] Poly(N-isopropylacrylamide)-based Smart Surfaces for Cell Sheet Tissue Engineering. Sigma-Aldrich, n.d. http://www.sigmaaldrich.com/technical-documents/articles/material-matters/poly-n-isopropylacrylamide.html.

[27] Hsieh F-Y, Lin H-H, Hsu S. 3D bioprinting of neural stem cell-laden thermoresponsive biodegradable polyurethane hydrogel and potential in central nervous system repair. Biomaterials 2015;71:48–57. https://doi.org/10.1016/j.biomaterials.2015.08.028.

[28] Kesti M, Müller M, Becher J, Schnabelrauch M, D'Este M, Eglin D, et al. A versatile bioink for three-dimensional printing of cellular scaffolds based on thermally and photo-triggered tandem gelation. Acta Biomater 2015;11:162–72. https://doi.org/10.1016/j.actbio.2014.09.033.

[29] Symposium CF. Silicon biochemistry. John Wiley & Sons; 2008.

[30] Thornton PD, Billah SMR, Cameron NR. Enzyme-degradable self-assembled hydrogels from polyalanine-modified poly(ethylene glycol) star polymers. Macromol Rapid Commun 2013;34:257–62. https://doi.org/10.1002/marc.201200649.

[31] Fahlgren A, Bratengeier C, Gelmi A, Semeins CM, Klein-Nulend J, Jager EWH, et al. Biocompatibility of polypyrrole with human primary osteoblasts and the effect of dopants. PLoS One 2015;10:e0134023. https://doi.org/10.1371/journal.pone.0134023.

[32] Kim. Effect of dopant mixture on the conductivity and thermal stability of polyaniline/nomex conductive fabric. J Appl Polym Sci 2001. Wiley Online Library n.d. http://onlinelibrary.wiley.com/doi/10.1002/app.10211/full.

[33] Varela-Álvarez A, Sordo JA, Scuseria GE. Doping of polyaniline by Acid–Base Chemistry: density functional calculations with periodic boundary conditions. J Am Chem Soc 2005;127:11318–27. https://doi.org/10.1021/ja051012t.

[34] Chen M-C, Sun Y-C, Chen Y-H. Electrically conductive nanofibers with highly oriented structures and their potential application in skeletal muscle tissue engineering. Acta Biomater 2013;9:5562–72. https://doi.org/10.1016/j.actbio.2012.10.024.

[35] Cho Y, Borgens RB. Biotin-doped porous polypyrrole films for electrically controlled nanoparticle release. Langmuir 2011;27:6316–22. https://doi.org/10.1021/la200160q.

[36] Hsiao C-W, Bai M-Y, Chang Y, Chung M-F, Lee T-Y, Wu C-T, et al. Electrical coupling of isolated cardiomyocyte clusters grown on aligned conductive nanofibrous meshes for their synchronized beating. Biomaterials 2013;34:1063–72. https://doi.org/10.1016/j.biomaterials.2012.10.065.

[37] Tandon N, Cannizzaro C, Chao P-HG, Maidhof R, Marsano A, Au HTH, et al. Electrical stimulation systems for cardiac tissue engineering. Nat Protoc 2009;4:155−73. https://doi.org/10.1038/nprot.2008.183.

[38] Shin SR, Li Y-C, Jang HL, Khoshakhlagh P, Akbari M, Nasajpour A, et al. Graphene-based materials for tissue engineering. Adv Drug Deliv Rev 2016;105:255−74. https://doi.org/10.1016/j.addr.2016.03.007.

[39] Durmus NG, Tasoglu S, Demirci U. Bioprinting: functional droplet networks. Nat Mater 2013;12:478−9. https://doi.org/10.1038/nmat3665.

[40] Tan JL, Tien J, Pirone DM, Gray DS, Bhadriraju K, Chen CS. Cells lying on a bed of microneedles: an approach to isolate mechanical force. Proc Natl Acad Sci 2003;100:1484−9. https://doi.org/10.1073/pnas.0235407100.

[41] Novosel EC, Kleinhans C, Kluger PJ. Vascularization is the key challenge in tissue engineering. Adv Drug Deliv Rev 2011;63:300−11. https://doi.org/10.1016/j.addr.2011.03.004.

[42] Jain RK, Au P, Tam J, Duda DG, Fukumura D. Engineering vascularized tissue. Nat Biotechnol 2005;23:821−3. https://doi.org/10.1038/nbt0705-821.

[43] Rouwkema J, Rivron NC, van Blitterswijk CA. Vascularization in tissue engineering. Trends Biotechnol 2008;26:434−41. https://doi.org/10.1016/j.tibtech.2008.04.009.

[44] Paulsen SJ, Miller JS. Tissue vascularization through 3D printing: will technology bring us flow? Dev Dyn 2015;244:629−40. https://doi.org/10.1002/dvdy.24254.

[45] Hinton TJ, Jallerat Q, Palchesko RN, Park JH, Grodzicki MS, Shue H-J, et al. Three-dimensional printing of complex biological structures by freeform reversible embedding of suspended hydrogels. Sci Adv 2015;1:e1500758. https://doi.org/10.1126/sciadv.1500758.

[46] Feinberg AW, Feigel A, Shevkoplyas SS, Sheehy S, Whitesides GM, Parker KK. Muscular thin films for building actuators and powering devices. Science 2007;317:1366−70. https://doi.org/10.1126/science.1146885.

[47] Boudou T, Legant WR, Mu A, Borochin MA, Thavandiran N, Radisic M, et al. A microfabricated platform to measure and manipulate the mechanics of engineered cardiac microtissues. Tissue Eng Part A 2012;18:910−9. https://doi.org/10.1089/ten.tea.2011.0341.

[48] Miller JS, Stevens KR, Yang MT, Baker BM, Nguyen D-HT, Cohen DM, et al. Rapid casting of patterned vascular networks for perfusable engineered three-dimensional tissues. Nat Mater 2012;11:768−74. https://doi.org/10.1038/nmat3357.

[49] Lind JU, Busbee TA, Valentine AD, Pasqualini FS, Yuan H, Yadid M, et al. Instrumented cardiac microphysiological devices via multimaterial three-dimensional printing. Nat Mater 2017;16:303−8. https://doi.org/10.1038/nmat4782.

[50] Park S-J, Gazzola M, Park KS, Park S, Santo VD, Blevins EL, et al. Phototactic guidance of a tissue-engineered soft-robotic ray. Science 2016;353:158−62. https://doi.org/10.1126/science.aaf4292.

FURTHER READING

[1] Momeni F, Mehdi M, Hassani NS, Liu X, Ni J. A review of 4D printing. Mater Des 2017;122:42−79. https://doi.org/10.1016/j.matdes.2017.02.068.

Bioprinting Cardiovascular Organs

Yasin Hussain[1,2], Jonathan T. Butcher[3]

Department of Radiology, Weill Cornell Medicine, New York, NY, United States[1]; Dalio Institute of Cardiovascular Imaging, NewYork-Presbyterian Hospital, New York, NY, United States[2]; Biomedical Engineering, Cornell University, Ithaca, NY, United States[3]

INTRODUCTION

The heart is a complex organ that is indispensible for sustained human life. The heart contains several fully differentiated cell lines that make up multiple structures including valves, coronary arteries, and myocardial tissue. Diseases of cardiac etiology are the leading cause of morbidity and mortality [1]. The available options in medically refractory conditions include replacement of a malfunctioning structure, which could include coronary artery bypass grafts, to total valve replacement and heart transplantation. Replacement grafts can include autografts, allografts, xenografts, and artificial prostheses. Each graft can be associated with intrinsic limitations, which could include shortage of tissue, rejection, thrombogenesis, and limited durability [2]. Bioprinting has emerged as a branch of tissue engineering which uses biodegradable polymers to build complex tissue constructs currently being used as surgical planning models [3]. The technique utilizes an individual's own radiographic images into which the scaffold is introduced to formulate tissue constructs using living cells and biological components [4] (Fig. 9.1). Bioprinting enables accurate placement of cells, extracellular matrices, and other biomolecules to replicate native tissue [5]. It can form mechanically heterogeneous structures, which enable scientists and engineers to move one step closer to replicating native tissue [6].

BIOPRINTING TECHNOLOGIES

Bioprinting techniques, similar to 3D printing, deposit 2D layers that are of single cell thickness in order to form the desired shape [7]. An individual's medical images (i.e., computed tomography [CT], magnetic resonance imaging, or echocardiography) can be used to formulate the desired 3D structure [8]. The building blocks used in this process are bioinks that contain cells, components of extracellular matrix (ECM), and other biomaterials. This technique allows the 3D scaffolds to be anatomically precise and physiologically relevant. Bioprinting currently uses inkjet-based, pressure-assisted, laser-assisted techniques and stereolithography, all of which are reviewed below (Fig. 9.2) [9].

3D Printing Applications in Cardiovascular Medicine. https://doi.org/10.1016/B978-0-12-803917-5.00009-2

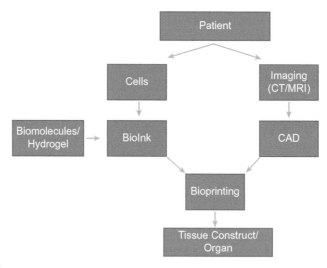

FIGURE 9.1

Schema of the bioprinting process. Tissue and imaging are obtained from a patient. Tissue is used to form the bioink, while radiographic images are used with the aid of computer-aided design to form a computerized model. The data is inputted into a bioprinter, which produces the desired construct.

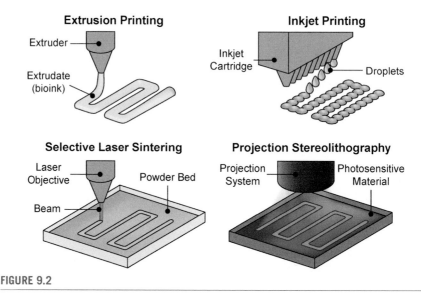

FIGURE 9.2

Commonly used bioprinting technologies.

Reproduced with permission from Miller JS, Burdick JA. Editorial: special issue on 3D printing of biomaterials.
ACS Biomater Sci Eng 2016;2:1658–61.

INKJET-BASED BIOPRINTING

Inkjet-based bioprinting is a noncontact printing process similar to the conventional desktop inkjet printers, which deposit ink droplets on paper. In this case, however, bioink droplets are deposited onto a hydrogel substrate using a computer software [10]. The two main types of inkjet bioprinting are thermal and piezoelectric bioprinting. Heating is used to generate the ink droplets in the thermal method, wherein hot bubbles force the ink out of the nozzle and onto the hydrogel. Although inkjet-based bioprinting uses very high temperatures, the exposure lasts a few microseconds; hence, the temperature rise of ejected mammalian cells is about 10°C above their surrounding which results in average cell viability of 90% [11]. The viscosity of the ink used, the temperature gradient, and the force of the pulse generated are all features affecting the size of the droplets being produced (with volumes ranging from 10 to 150 pL) [12].

The piezoelectric method uses piezoelectric actuators; stacks of thin ceramic layers that extend when voltage is applied and convert electrical energy into linear high-speed motion resulting in transient pressures which generate droplets that are regular and equal in size [13,14]. The frequencies used by the piezoelectric method can at times cause damage to the cell membrane leading to cell lysis [12]. Despite that risk, mammalian cells such as human osteoblasts and fibroblasts have been deposited using the piezoelectric method with 90% viability [14]. Advantages of inkjet bioprinting include low cost, high speed, and production of high resolution constructs with the ability to use multiple cell types; disadvantages include the thermal and mechanical stress affecting viability, limitation to liquid-based materials, weak mechanical properties of the constructs, and insufficient time to produce tissue of clinically sized thickness [5,15,16].

PRESSURE/EXTRUSION-BASED BIOPRINTING

Extrusion-based bioprinting uses high pressures to expel bioinks in filaments or fibers to form a layer-by-layer tissue scaffold [17]. The materials are extruded through a microscale nozzle in a continuous fashion to form the desired product. The nozzle is moved by a robotic arm along an $X-Y$ axis while the material is dispensed. The dispenser can also be moved vertically along a Z axis to allow for the formation of three-dimensional structures as layers deposited onto the substrate [18]. This bioprinting method utilizes three different deposition mechanisms, namely pneumatic, piston, and screw-based techniques. The pneumatic-based approach drives the biomaterials out of the nozzle utilizing pressurized air at a controlled volume flow rate. The viscosity of the bioink influences the accuracy of this technique as flow rates can vary depending on the properties of biomaterial used [19]. The piston- and screw-based techniques mechanically force the biomaterials out of the nozzle. These techniques use large forces to control the volume of solution that is being extruded. The downside to using such large driving forces is the fact that they can cause rupture of cellular membranes and cell lysis [20].

Increasing the diameter of the nozzle or utilizing viscous materials can dispense thicker filaments which produce mechanically stable structures [21]. A balance needs to be struck, however, since thicker filaments produce materials with low porosity which inevitably reduces cellular diffusion and limits the cellular functionality of the construct [22]. Pressure-based bioprinting can use numerous materials including: hydrogels, cells, and proteins with dispersions of low and high viscosity. Overall, the technique is slow, of moderate cost, produces low resolution, and requires specific matching of the material utilized with the implanting matrix to preserve viability [23,24].

LASER-ASSISTED BIOPRINTING

This technique uses laser as its energy source which is of ultraviolet (UV) or near UV wavelengths [5–25]. The setup of the laser bioprinter consists of two parallel glass slides. A laser radiation absorbing metallic material (i.e., gold) is part of the top slide, which is covered by the liquid biomaterials. The energy from the laser causes the biomaterials to evaporate and drop to the bottom slide, which contains a cell culture medium to create adhesions and sustain the assembly as the structure is being built [5]. This is a nozzle-free technology, which differentiates it from the prior described techniques [26]. The absence of an orifice avoids clogging, which limits the cellular damage that occurs with other printing modalities [27]. This technique is versatile in terms of biomaterials used and their viscosities which include hydrogels, cells, and proteins to produce structures of extremely high viability (nearing 100%) [28,29] and resolutions greater than 20 μm [26,30,31]. The viscosity and thickness of the material on the metallic film, the absorbability of the substrate, and the energy of the laser used are all variables that influence the resolution of the product [32]. Laser-assisted bioprinting is more complex, expensive, and timely than the other techniques. In order to construct complex structures with heterogeneous components, multiple metallic films would be required with each film coated with the required individual biomaterial [33].

STEREOLITHOGRAPHY

Stereolithography is a solid free-form technology that uses a photosensitive liquid polymer and solidifies it using light [34]. In this nozzle-free technique, the light's intensity is controlled using digital arrays to polymerize and cross-link the liquid polymer biomaterials [35]. This technique has the highest fabrication accuracy. The 3D structure is built layer by layer as the light cross-links each layer to geometrically solidify the structure [36], with timing of production depending on size and thickness of the structure [37]. The two different methods to form the 2D layers using stereolithography are beam scanning and mask-image projection technique. In the beam scanning technique, the laser beam scans the photocurable bioink to solidify a 2D layer [38]. The type of bioink, the speed of scanning, and the power of the laser used contribute to the mechanical properties of the construct [39]. In the mask-image projection technique, a predefined image is projected onto the photocurable bioink and the 2D layer is formed simultaneously [40]. This system can be much faster than the beam scanning technique. This technique is currently restricted by the limited numbers of photosensitive polymers available, the lack of biodegradability, and compatibility [41]. Another disadvantage is the harmful effect from residual toxic photocured material as there are free radicals that form from this light-induced chemical reaction which can damage cells and proteins [42].

BIOINKS

Designing the ideal bioink is a complex task, as multiple materialistic and biological parameters must be addressed to produce a functioning structure. Bioink may be in liquid or solid form to shape the desired product [5]. Essential biological properties include compatibility and biological activity. Material parameters that should be considered include viscosity, surface tension, and cross-linking properties; all of which are essential for mechanics and functionality of the construct [43]. Highly viscous bioinks are more likely to form more rigid structures. On the other hand, higher viscosity will limit the ease of flow of the bioink; thus requiring printing processes that allow for higher pressures. Bioinks with adjustable viscosities are generally preferred as they can be used in different bioprinting methods. Inkjet bioprinters are capable of printing bioinks of viscosity values 10 mPa s [44]; pressure-based bioprinters use materials with viscosities ranging from $3-6 \times 10^7$ mPa s [45], and laser printers use bioinks with viscosities ranging from 1 to 300 mPa s [32]. The rigidity and strength of the structure influences cellular behavior as does the degradability of the construct if it were to be replaced by cells and an ECM [43]. Mechanical properties of the bioinks can be enhanced by incorporating nanoparticles [46]. The functionality is dictated by the cellular viability and biocompatibility which allows for cellular adhesion and differentiation [43]. The most commonly used bioinks will be discussed in this section.

CELL-LADEN HYDROGELS

The ability to use cell-laden hydrogels in inkjet-based, pressure-assisted, and laser-based bioprinting and their capacity to mimic cellular milieu have made them the most commonly used bioinks [47]. There are both natural and synthetic cell-laden hydrogel formulations. Natural biomaterials such as fibrin, collagen, gelatin, and hyaluronic acid offer bioactivity by their innate nature and can resemble the ECM [43]. Alginate, another natural hydrogel, is used in both inkjet-based and pressure-assisted bioprinting [43]. This material does not offer bioactivity and is often mixed and used with other natural hydrogels to promote cellular adhesion and activity [48]. Although they offer high cell survivability, the natural polymers are often weak and unsustainable [5]. Hence, the majority of natural hydrogels are used as components in a mixture to form the desired qualities of a bioink [43]. Natural polymers have backbone side chains that can be manipulated and induced to alter their property [49]. Some hydrogels like fibrin get structural enhancement by the use of cross-linkers to achieve the necessary tensile strength [4,5]. Hyaluronic acid, for example, has been manipulated to cross-link to form supramolecular hydrogels that are capable of altering their viscosity under pressure and healing the structure after the pressure is resolved [50], making it an attractive choice for pressure-based bioprinting.

Synthetic hydrogels lack the bioactivity of the natural alternates, but offer more adjustable mechanical properties [43]. Pluronic and polyethylene glycol (PEG) are both commonly used synthetic hydrogels. Pluronic lacks stability and has been observed to erode after printing, and as a result its use has been limited to that of supporting material [51]. PEG can be used as bioink in laser-based, inkjet and pressure-assisted bioprinting [52,53]. Synthetic hydrogels are cross-linked using UV light exposure [5—13]. Hence synthetic hydrogels can be used to form structurally stable constructs [43]; however, they are hydrophilic compounds that lack the ability for cellular attachment and activity. Therefore, they are often mixed with natural hydrogels or manipulated chemically to allow them to become more functional [43].

CELL SUSPENSION

Cell Suspension is a type of bioink in which single cells or aggregates of cells multiply as they lie suspended in a predefined cell media. This bioink has been used in inkjet printers [54], and the cells can be used to aggregate without the need for a scaffold [55,56]. The cell-cell interactions allow the bioink to self-assemble without the need of a support layer [57,58]. Multicellular bioinks used have been shown to undergo cellular fusion after printing leading to the formation of vascular constructs. This bioink has also been used to develop liver models using pressure-assisted bioprinting [59].

DECELLULARIZED EXTRACELLULAR MATRIX

This bioink utilizes the extracellular matrix (ECM) as its main component. The tissue of interest is stripped of its cells with preservation of the ECM, which is then dissolved in a solution to form the bioink [43]. Modifications to this bioink are mostly controlled by the polymer carrying it. The choice of polymer can adjust the viscosity, solubility, and bioactivity of the ink [60]. The decellularization process is costly and intensive as it involves a multistep approach [43]. Using this approach, cardiac tissue composed of multiple tissue and cell lines can be manufactured [61]. Liver ECM has also been used in bioprinting and has been shown to enhance activity of stem cells and other hepatocyte cell lines [62].

BIOMOLECULES

Growth factors have been used in 2D systems and have been shown to have effects on cellular differentiation [63]. These biomolecules can be introduced into the bioink by mixing or through functional attachment to the gel being used [64,65]. These biomolecules are sensitive and unfavorable environments should be avoided when possible [66,67]. Nevertheless, biomolecules influence cellular differentiation when used within the substrate prior to bioprinting or when added onto the printed construct [68].

BIOPRINTING CARDIOVASCULAR TISSUE
CARDIAC VALVES

Of the four heart valves, the aortic valve is the one that is most commonly involved in heart disease, with either stenosis or valvular insufficiency. The aortic valve has been the focus of the surgical and interventional fields in terms of repair and replacement therapies; living grafts are especially useful in the young who need growth and biointegration. The Ross procedure or pulmonary autograft, replaces the aortic valve with the patient's healthy pulmonary valve, and places a mechanical/bioprosthetic valve in the hemodynamically weak pulmonary position [69]. There have been some concerns regarding the durability of the pulmonary replacement in the long term [70]. Tissue engineered heart valves (TEHV) are the ideal living valve replacement, especially in the pediatric population. Multiple techniques are available for TEHVs including decellularization, mold suturing, and bioprinting [69]. Bioprinting and the attempts that have been made thus far will be discussed in this section.

AORTIC VALVE STRUCTURE AND BIOMECHANICS

The aortic valve, from the periphery to the center, consists of the aortic root, the annulus connecting the valve root to the cusps, and the commissures between the three valve leaflets [71]. The root consists of the adventitia, media, and intima layers; it is made up of endothelial cells, smooth muscle cells, and fibroblasts. The cusps are also three-layered consisting of stratified connective tissue namely the fibrosa, spongiosa (middle layer), and ventricularis [72]. Collagen fibers make up the fibrosa layer which takes on the cusps' workload [73], while the ventricularis layer is responsible for the expansion and convolution of the cusps during the cardiac cycle [5]. The cusps' structure permits accommodation of significant pressure that occurs throughout the cardiac cycle, pressures that can range from 50 to 500 kPa [74]. The valve leaflets' tensile strength reaches up to 10 times that of the physiologic pressure they endure, which contributes to durability as aortic valve leaflets are constantly facing high pressures during systole and diastole [75]. The aortic root has been shown to deform by 5%

longitudinally and circumferentially due to the applied pressures [76]. The cusps are lined by endothelial cells surrounding an inner layer of interstitial cells [77]. The valve leaflets' interstitial cells are of fibroblast phenotype [78,79]. They are sensitive to the transvalvular pressures and damage to the leaflets eventually results in collagen deposition by the fibroblasts and a phenotypic transformation to myofibroblasts leading to valvular contraction and calcification, which manifests clinically as aortic stenosis [80].

AORTIC VALVE CONSTRUCTS: PARAMETERS FOR BIOPRINTING

The native valve structure is specifically engineered to withstand the physiologic transvalvular pressures. Long-term stability and longevity of the aortic valve constructs, their ability to cellularize and remodel, and prevention of regurgitation are issues that need to be addressed in valvular construct design, as inaccuracies can lead to dysfunctional constructs.

Tissue engineering of aortic valves has been attempted with decellularized valves from allogenic/xenogenic sources and with contrived synthetic constructs [81]. The decellularized valves retain their ECM and the original valve structure which makes for better tissue but are antigenic compared to synthetic material, especially xenogenic valves which are much more common than allogenic valves [82]. With synthetic valves there is no concern regarding immunogenicity and there is a lower risk of thrombogenesis, calcification, and corrosion compared to polymeric valves; however, it is difficult to manipulate synthetic material to form a mechanically accurate valve [83].

Attempts to create a functioning valve by the above-mentioned tissue engineering methods have been difficult. The native valves' specific geometry and heterogeneity are difficult to re-create [5]. Bioprinting allows for larger diversity in the structure of the constructs and is thought to potentially overcome some of the hurdles faced by other tissue engineering techniques. Several parameters should be considered in the engineering and preplanning processes of bioprinting aortic valves. The dimensions and geometry of the valve should be as accurate as possible; with the help of advanced imaging, accurate measurements should be taken to aid in construct design [84]. Inaccurate dimensions would lead to uneven stresses on the valve causing damage, which could lead to calcification of the valve [5]. It is prudent to note that the aortic valve leaflet fibrils are morphologically anisotropic, meaning, by design the fibers are aligned in multiple directions. Synthetic valve design lays the fibers in one direction, which increases the risk of valvular damage in the face of transvalvular pressures [81].

Another native valve property that should not be overlooked is the heterogeneity in its thickness [85]. There is variation in thickness between the different cusps, and even within the same cusp, in order to maximize valve functionality. With the help of imaging techniques, thicknesses should be determined and if possible re-created. Homogeneity in the cusps' thickness could lead to increased stress on the valve as a unit and cause damage [5]. Finally, the cellular components that make up the valve should also be considered when bioprinting and formulating the bioink for pre- or postcellularization of the valve. The valvular interstitial cells give the valve its structure, allowing for growth, remodeling, and repair [5]. Endothelial cells cover the surface of the valve and are essential for reducing the risk of thrombosis [5].

EXAMPLES OF BIOPRINTING AORTIC VALVES

Bioprinting has elicited significant interest among biomedical researchers and tissue engineers; yet, only a limited number of laboratories have experimented with aortic valve bioprinting. In this section, currently available data will be reviewed.

Researchers at Cornell University have been successful in bioprinting a porcine aortic valve [16]. The porcine aortic valve was fixed in formalin and scanned using micro CT imaging (Fig. 9.3A). The scanned images were then extracted into 3D geometries using standard tessellated language (STL) files. The base material of the hydrogel used was poly(ethylene glycol)-diacrylate (PEG-DA), which was chosen for its ease of modification [16]. Two different molecular weights were used, 700 and 8000, to assess for differences in stiffness. A photoinitiator (Irgacure 2959) was used to enable cross-linking when UV light was used. The ratio of the concentration of photoinitiator to PEG-DA remained stable at 1%. To achieve the desired viscosity during extrusion, alginate was added to the mix at 10%−15% weight/volume. An extrusion-based bioprinter was used, which was linked to the

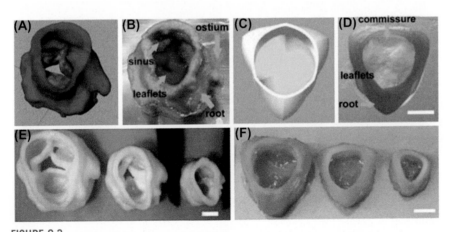

FIGURE 9.3

(A) CT imaging of porcine aortic valve fixed in formalin. (B) Bioprinted valve, with root made using 700 MW PEG-DA hydrogel and leaflets using 700/8000 MW PEG-DA. (C) Bioprinted valves with 700 MW PEG-DA at different inner diameters 22, 16 and 12 mm. (D) Axisymmetric model. (E) Axisymmetric valve model printed using blend of hydrogels. (F) Printed with different internal diameter 22, 17, and 12 mm.

Reproduced with permission from Hockaday LA, et al. Rapid 3D printing of anatomically accurate and mechanically heterogeneous aortic valve hydrogel scaffolds. Biofabrication 2012;4:035005.

UV-LED source for simultaneous printing and cross-linking. With variability in molecular weight, cross-linker concentration, cross-linking time, researchers were able to print valves with variable stiffness (5.3—74.6 kPa). The technique used enabled heterogeneity of the valve with different stiffer formulation for the aortic root and a more extensible formulation for the leaflets (Fig. 9.3B). After printing, the valve's geometric fidelity was assessed by micro-CT quantitative analysis, which showed high precision that is reduced when the valve size is reduced (valves from 12 to 22 mm inner diameter were printed) (Fig. 9.3C). To test cellular viability, constructs were stained with calcein—AM and ethidium homodimer (Live/Dead kit) and then imaged using epifluorescence stereomicroscope. Valvular interstitial cells were seeded onto the printed structure and the viability was 100% at 21 days (Fig. 9.4A and B). The investigators were able to fabricate a structurally accurate scaffold that was heterogeneous and had the ability to allow cell growth using bioprinting [16].

The same investigators were successful in printing an aortic valve using different biomaterials [86]. A porcine aortic valve was preserved in formalin and scanned with micro-CT and the images were used to form an STL file for bioprinting. The bioink used was a mixture of gelatin and alginate with phosphate buffered saline

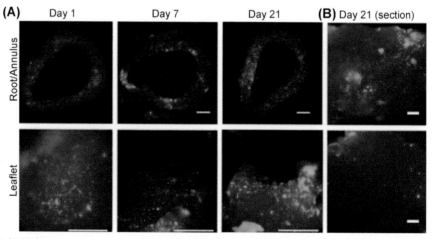

FIGURE 9.4

Constructs were stained with Calcein-AM and ethidium homodimer (Live-Dead, Invitrogen) and imaged using epifluorescence stereomicroscope. (A) Live Porcine Aortic Valve Interstitial Cells (PAVIC) visible on surface of annulus (top) and leaflets (bottom) on days 1, 7, and 21. (B) PAIVCs were detected in the interstitium of annulus and leaflet up to 21 days post scaffold printing.

Reproduced with permission from Hockaday LA, et al. Rapid 3D printing of anatomically accurate and mechanically heterogeneous aortic valve hydrogel scaffolds. Biofabrication 2012;4:035005.

as the solvent. Valve interstitial cells cultured from a porcine aortic valve and smooth muscle cells isolated from human aortic root were both mixed into the bioink separately to print aortic valve leaflet and aortic root sinus, respectively. An extrusion-based printer was used to formulate the structures based on the STL files which were then ionically cross-linked in calcium chloride, since the calcium ions bind the alginate molecules to cross-link the gel. Hydrogels encapsulated with cells had stable stiffness over time likely due to deposition of ECM; however, acellular hydrogels would soften as the calcium ions dissolve into the substrate [5]. The printed valve exhibited geometry similar to the derived images from the native valve and fluorescent labeling verified the heterogeneity of cells in the structure. This study highlighted the ability to produce aortic valves with cellular heterogeneity and high cellular viability. Duan et al. later used methacrylated hyaluronic acid and methacrylated gelatin hydrogel mix to print a trileaflet structure [86]. Similar to the previous methods, CT imaging was used to create an STL file that was eventually printed using an extrusion-based printer. An acellular hydrogel was used for the aortic root and the valve leaflets were printed using a cell-encapsulated hydrogel. The stiffness of the structure ranged from (13−4.2 kPa) based on the concentration of methacrylated gelatin. Cellular viability was >90% at 7 days and production of ECM was noted. The structures with lower stiffness allowed the interstitial cells to spread and larger amounts of ECM were deposited showing higher cellular activity. The addition of methylated gelatin to the methylated hyaluronic acid increased cell proliferation, migration, and spreading. These findings were significant as they highlight the importance of hydrogel manipulation to regulate cellular activity and spread [86].

LIMITATIONS

The aim of bioprinting a cardiac valve is to produce a product with identical features to the native aortic valve. That entails high resemblance in geometry and mechanical heterogeneity similar to that of the native valve. Hydrogels used thus far cannot withstand the physiologic pressures due to their pliability [86]. Cross-linking can stiffen the hydrogels [5]; however, that comes at a cost. It can over stiffen the hydrogel which then hinders cellular growth and migration. Chemical cross-linking with the use of UV rays can also cause damage to cells and to extracellular proteins. Structural integrity of hydrogels can be preserved with periodic exposure to chemical of UV cross-linking [65], which is not possible under physiologic conditions. Endothelialisation of the valve has not been attempted to date; in order to reduce the risk of thrombogenesis, the leaflets' interstitial cell layer should be covered with an endothelial cell layer [5]. Though bioprinting has initiated the process of creating aortic valve constructs that are structurally similar to native valves, this route is hindered by the biomaterials being used and limited printing methodologies, but will surely continue to evolve as engineers and scientists continue to develop innovative bioprinting methodologies.

VASCULATURE

Just like native tissue, artificially engineered tissue requires adequate oxygen and nutrient delivery, as well as waste removal. Vascular constructs can help maintain the required blood flow to allow for metabolic functioning of tissue [87,88]. Tissue engineering will remain limited without adequate vascular constructs to allow for the above-mentioned exchanges to help retain cellular viability; this will be especially important for larger tissue constructs such as liver, pancreatic and cardiac tissue. Cells undergo apoptosis without a continuous supply of blood, which would occur if the nearest capillary is further than 200 µm away [87]. A viable vascular network is pivotal for tissue formation and growth [89]. Bioengineered tissue with thicknesses greater than 1 mm cannot retain viability without vascularization [90].

Vascular grafts can be useful for many patients with diseases involving vessel stenosis such as coronary artery disease, cerebrovascular disease, and peripheral vascular disease that often require revascularization. Synthetic constructs of large or medium diameter have shown satisfactory results; smaller caliber vascular constructs (<6 mm in diameter), however, remain a suboptimal replacement option to autologous grafts [91]. This has largely been due to synthetic material surface properties, failure to grow a healthy endothelial cell layer, and lack of vessel compliance [92,93]. Bioprinting is an attractive option to overcome this barrier and enables the engineering of small-diameter endothelialized blood vessels.

STRUCTURE AND FUNCTION

Vasculature is extremely complex and variable with multiple ranges in size and layers existing in native blood vessels. The inner diameter ranges from 10 µm in the smallest capillaries to 30 mm in the largest arteries [94]. Mechanically, vessels are also very variable in the amount of pressure they can withstand ranging from 5 to 120 mm Hg and the speed of blood flow which ranges from 0.04 to 40 cm/s. The mechanical variability is related to the multiple layers and the building blocks that a vessel can contain. Larger arteries are made up of multiple layers; an intima, media, and adventitial layer from the lumen to the periphery [95]. The intima is a monolayer of endothelial cells attached to the basement membrane [96]. The media's main building blocks are smooth muscle cells and collagenous fibers. The adventitia, the outermost layer, is composed of collagen fibers, ECM, and fibroblasts [97]. Smaller vessels such as capillaries can withstand lower pressures because they are monolayered with an endothelial cell layer. Consideration of the desired vessel function and acquiring the adequate geometry is essential for bioprinting to formulate mechanically relevant vessels.

BIOPRINTING VASCULAR CONSTRUCTS

There are two main approaches to bioprinting blood vessels, a scaffold-based and a scaffold-free approach [91]. In the former approach, cells are mixed into the hydrogels or ECM and printed in exogenous biomaterial that resembles the vessel with the desired caliber. In the scaffold-free approach, there is no structural support and the cells form tissues via self-assembly [98].

SCAFFOLD-BASED INDIRECT VESSEL BIOPRINTING

In this method of vascular construct bioprinting, a sacrificial material is used to generate the desired network of vessels. The sacrificial hydrogel (i.e., alginate, gelatin, carbohydrate, Pluronic F127) forms the construct and is removed thermally or chemically after printing [99]. Indirect vessel printing using pressure-assisted bioprinting of agarose to form the sacrificial template has been successful [100]. Embedding the hydrogel with microchannels increases cellular viability to 90% as shown when using methacrylated gelatin and PEG [100]. Indirect pressure-assisted bioprinting of vessels has also been shown with the use of Pluronic F127 (1—47), a nonionic surfactant polyol, as a sacrificial ink that is easily removed. Furthermore, vascular networks have been successfully printed using fibroblasts in methacrylated gelatin. Inkjet-based bioprinting has been used for indirect vessel bioprinting. The scaffold used is sacrificial gelatin hydrogel onto which layers of cell-laden hydrogel are printed. The sacrificial material is then thermally removed and the channels are vascularized with endothelial cells [91]. A layer-by-layer printing approach has been described, in which collagen is the first layer followed by solidified gelatin [101]. The bioink used after that step is a mixture of gelatin and human umbilical vein endothelial cells (HUVECs). The gelatin is liquefied when incubated at 37°C. The construct forms a layer of endothelium and supports cellular viability under physiologic flow conditions [101].

SCAFFOLD-BASED DIRECT VESSEL BIOPRINTING

The vascular construct's shape and size is predetermined; a cell-laden hydrogel is directly printed into the desired structure and solidified after printing into a cellular microenvironment. The cells proliferate and form rigid constructs after remodeling [91]. Cellular viability in this direct printing technique is highly reliable on mechanical properties of the bioink [102]. Pressure-assisted extrusion-based bioprinting has been used to print vascular constructs around hepatocytes to form channels within the liver. In such instances, the bioink was a composite of gelatin, alginate, and fibrinogen [103]. Cross-linking agents were used to solidify the structure. Others used direct extrusion of PEG derivatives along with hyaluronic acid and gelatin which were cross-linked using multifunctionalized polymers [104]. Bioprinting of those constructs with 3T3 cells deposited into the tubular structures showed cell viability 4 weeks post manufacturing [104]. Another extrusion method that has been shown to work utilizes a coaxial nozzle with an outer and inner tube at the output side of the printer. As the bioink is expressed from the outermost nozzle, the cross-linking solution is expressed from the innermost nozzle and immediate cross-linking occurs to solidify the structure [105]. Thus far alginate has been used successfully due to its rapid cross-linking capacity when treated with calcium [106]. The addition of human umbilical vein smooth muscle cells (HUVSMCs) is performed by loading cells into the alginate prior to printing [107], with greater than 80% cellular viability at 7 days [108]. Investigators have made advances in this coaxial extrusion technique by adding variability to the flow rate and concentrations of alginate and calcium chloride which enhances fusion to adjacent hollow filaments.

Inkjet-based bioprinting has been used to form direct scaffold-based vascular constructs. Patterns have been produced in liquid media using hydrogels with entrapped hematopoietic stem cells that were cross-linked on demand [109]. Other investigators have utilized a fibrin hydrogel infused with human microvascular endothelial cells to form microvessel constructs [110]. After 21 days, there was confluence of cells in the lining and the integrity of the structure was maintained; branching tubular structures were also noted. Laser-based bioprinting has been used to form vascular constructs. A single layer of HUVECs was able to form interconnections within 1 day; however, branches formed were very weak and unstable [111]. The structure was enhanced with the addition of an outer layer of HUVSMCs [112]. Structural stability was observed with bioprinting Y-shaped tubes using mouse fibroblast-laden sodium alginate [113], with limited postprinting cellular bioavailability. A more successful technique used gold nanorods within the collagen cell infused hydrogel that absorb the laser beam and cause denaturation of collagen allowing for the formation of channels within the structure [114]. Endothelial cells could migrate throughout the channels, proliferate, and retain the 3D structure; cellular viability was reported at greater than 90% [114].

SCAFFOLD-FREE VESSEL BIOPRINTING

The interest in scaffold-free bioprinting of vascular constructs stems from some of the constraints associated with the scaffold-based designs. Yet, this technique remains plagued with limitations despite many years of research. Most important is the insufficient mechanical strength and slow or delayed cell growth due to the presence of residue from the used polymers [94,115]. Mechanical compatibility and tissue formation at a rate that matches the native tissue remains an unreached goal [91]. The advantage of this method is using the ECM from the cells within the bioink to formulate the construct, which is more biocompatible and forms a mechanical structure similar to the native vessel. For example, HUVEC coated human myofibroblasts were assembled into macrotissues and formed a vascular system when they were implanted into a chicken embryo [116]. Some investigators utilized multiple cell lines including HUVSMCs, human skin fibroblasts, and porcine aortic smooth muscle cells, which were placed into micropipettes to form cylinders which were sliced into rounded spheroids [58]. The spheroids were then expelled using pressure-assisted bioprinting alongside agarose mold and the spheroids were placed inside the mold to mature later on to a vascular structure, after which the mold was removed. The major limitation to this bioprinting approach has been the maturation time of the construct as the ECM takes time to deposit collagen and other fibrous proteins to attain the desired structure and tensile strength [91].

LIMITATIONS

The ability to ensure nutrient and oxygen delivery to bioengineered tissue is essential for tissue growth and the continued development of this field; hence the ability to form vascular constructs and vascularized tissue is vital. Bioprinting is unmatched by other engineering techniques in its ability to form anatomically and biologically accurate constructs; however, this field still faces considerable limitations that hinder its use at the clinical level. The choice of biomaterial, the stiffness of the construct, and the cellular viability post printing remain issues that need fine-tuning. Assuring stability of the vascular construct over a period of time and the ability to withstand the changing physiologic pressures and strains remain areas that need further testing and development. The ability to control the resolution and the spatial alignment formed by the channels as the cells migrate and rearrange is important going forward (especially in scaffold-free bioprinting). Vascular constructs should eventually re-assemble the entire vasculature from arteries, capillaries, and veins with different inner diameters and outer layers [91]. To date, this technology has failed in producing functioning monolayer capillary networks [91]. Many strides forward have been made in vascular tissue engineering using bioprinting technology; however, there are milestones that need to be achieved.

BIOPRINTING OF MYOCARDIAL TISSUE AND THE WHOLE HEART

Largely made of myocytes, the heart is responsible for pumping blood through the circulatory system [117]. Blood flows through its four chambers confined by its muscular walls and directed by the four valves within the heart [117]. Myocardial ischemia and/or infarction, which results from decreased or complete cessation of coronary blood flow to the myocardium, is the most common cause of death worldwide [118]. Depriving the cardiomyocytes of oxygen causes cell death, which is followed by a multistep inflammatory process leading to scar formation [119]. Cardiac myocytes in an adult are unable to divide and replicate; hence such injury causes loss of myocardial tissue [120]. The scar tissue, which is made up of fibroblasts, results in loss of ventricular function, and disruption of action potential propagation leads to abnormal electrical conduction. This section focuses on bioprinting developments when it comes to myocardial tissue and the whole heart.

STRUCTURE AND FUNCTION

The free ventricular and atrial walls of the heart consist of three layers, the endocardium, myocardium, and epicardium [117]. The endocardium is made up of a layer of endothelial cells, this layer acts to prevent thrombogenesis by keeping the blood from the other layers of the heart. The myocardium is the thickest layer and is made of cardiomyocytes; this layer is responsible for the primary contractile function. The muscles contract and relax forcing blood through the circulation and allowing the heart to fill during the cardiac cycle. The outermost layer, the pericardium, is composed of fibrous tissue and acts as an anchor for the heart in its surroundings [117].

BIOPRINTING MYOCARDIAL CONSTRUCTS

Loss of cardiac myocytes leads to loss of the heart's contractile function which can result in severe disease such as heart failure, and strategies to reverse and combat this disease are limited [121]. The heart's capacity to repair itself is extremely ineffective and the death of cardiomyocytes leads to scar tissue and limited contractile function [122]. Cellular implantation has been promising, but is limited by the viability of cells after injection [123]. Bioprinting myocardial tissue provides the required ECM for cells to grow and differentiate, thus providing the required structural support [124].

Investigators have been successful in producing functioning myocardial constructs using various bioprinting techniques. A group of investigators utilized isolated human fetal cardiomyocyte progenitor cells mixed into a sodium alginate gel and utilized calcium chloride to cross-link [125]. They used an inkjet-based bioprinter to form the construct which was 2×2 cm in dimensions. Cellular viability was assessed and was shown to be preserved at above 88% after 7 days; the authors were also able to show cellular migration within the alginate gel. The same group utilized the same cells and same printer to form a bioink using hyaluronic acid and gelatin mixture [126]. The cells retained their phenotype post printing, and subsequent attaching of the construct to a mouse model of myocardial infarction showed that the patch reduced left ventricular remodeling and improved myocardial viability, thus helping to preserve cardiac function. Other investigators have successfully used laser bioprinting to print human mesenchymal cells and endothelial cells onto a biodegradable polyester urethane urea patch [127]. The patch was engrafted into scar tissue from a myocardial infarction mouse model and showed enhanced angiogenesis and preserved cardiac function post infarction.

BIOPRINTING THE HEART

The heart is made up of many different cell types and different structures that allow it to perform its function. The valves, vascular tissue, and heart muscle all play essential roles in allowing blood to circulate. Investigators have been able to bioprint the heart's general structure. They utilized a 3D bioprinting technique that utilizes a thermoreversible hydrogel bath [99], a technique termed "free from reversible embedding of suspended hydrogels (FRESH)." This technique prints the construct into a hydrogel support bath, composed of gelatin micro particles, which significantly improves the structural fidelity. It allows a nozzle to move through the support bath and deposit the hydrogel forming the desired structure, while the temperature is subsequently adjusted to dissolve the support bath. To print a whole heart, a 5-day embryonic chick heart was utilized as a model. The heart was fixed and stained for heart cell nuclei, F-actin, and fibronectin to generate an image using confocal microscopy [99]. A 3D computed aided design (CAD) model was formulated using the confocal image and was found to be almost identical to the native structure, with less than 10% variability in internal and external anatomy, as shown in Fig. 9.5.

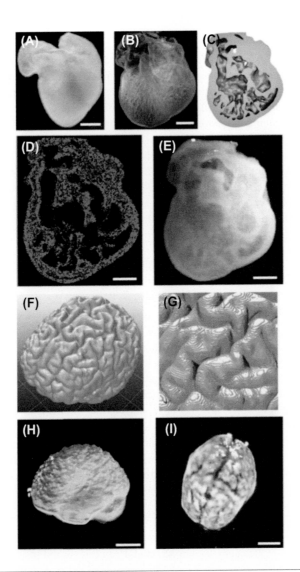

FIGURE 9.5

(A) Explanted embryonic chick heart. (B) 3D confocal microscopy image of 5-day-old chick heart showing fibronectin (*green*), nuclei (*blue*), and F-actin (*red*). (C) Computer-aided design (CAD) 3D model cross-section based on confocal image data showing complex internal structure. (D) 3D printed heart in cross-section in fluorescent alginate (*green*) re-creating the CAD model. (E) Dark-field image of printed heart showing internal structures. For comparison, 3D printing of the brain was performed in order to illustrate the complex architecture of the external white matter. (F) 3D model of human brain from magnetic resonance images. (G) Zoomed in view showing complexity of white matter folds. (H) Lateral view of 3D printed brain. (I) Top-down view of 3D printed brain.

Reproduced with permission from Hinton TJ, et al. Three-dimensional printing of complex biological structures by freeform reversible embedding of suspended hydrogels. Sci Adv 2015;1:e1500758.

LIMITATIONS

Bioprinting has advanced the field of tissue-engineered myocardial constructs and is rapidly approaching the ability to bioprint the entire heart; yet, there remain significant hurdles to clinical application. Further studies and testing is required to assess how the myocardial tissue constructs incorporate into natural tissue. The required electrical and vascular connections between the construct and the native heart tissue remain a challenge to clinical use. The viability of tissues and cells needs to be tested under physiologic stresses and for prolonged periods of time. As for bioprinting the entire heart, that target remains elusive. Vascularization on a multiscale level with multilevel networks remains out of reach. With the heart, there would need to be multiple bioinks and cell types printed and manipulated at once to create the vascular, muscular, endothelial, and valvular layers.

FUTURE DIRECTIONS

Bioprinting technology has been promising thus far. It has the potential to become an integral part of medical practice in the future as it enables fabrication of patient-specific scaffolds and tissues which are highly complex and convoluted. Various bioinks exist with multiple variations in cell type, biomolecules, and materials used to formulate the gel. Time and more extensive experience will be required in order to eventually develop an understanding for the optimal bioink formulation for each specific tissue construct. Studies regarding reversible cross-linking mechanisms and durability are the next steps going forward in bioink designs [128]. High resolution bioprinters and new techniques in printing such as the FRESH technique should be implemented to enhance the resolution and integrity of the constructs [99]. This niche of tissue engineering is very promising and will enable us to take strides along the path to tissue regeneration.

REFERENCES

[1] Mozaffarian D, Benjamin EJ, Go AS, et al. Heart disease and stroke statistics-2016 update: a report from the American Heart Association. Circulation 2016;133: e38−360.
[2] Bouten CV, Dankers PY, Driessen-Mol A, Pedron S, Brizard AM, Baaijens FP. Substrates for cardiovascular tissue engineering. Adv Drug Deliv Rev 2011;63: 221−41.
[3] Peltola SM, Melchels FP, Grijpma DW, Kellomaki M. A review of rapid prototyping techniques for tissue engineering purposes. Ann Med 2008;40:268−80.
[4] Guillemot F, Mironov V, Nakamura M. Bioprinting is coming of age: report from the international conference on bioprinting and biofabrication in Bordeaux (3B'09). Biofabrication 2010;2:010201.

[5] Jana S, Lerman A. Bioprinting a cardiac valve. Biotechnol Adv 2015;33:1503−21.

[6] Campbell PG, Weiss LE. Tissue engineering with the aid of inkjet printers. Expet Opin Biol Ther 2007;7:1123−7.

[7] Derby B. Printing and prototyping of tissues and scaffolds. Science (New York, NY) 2012;338:921−6.

[8] Ozbolat IT. Bioprinting scale-up tissue and organ constructs for transplantation. Trends Biotechnol 2015;33:395−400.

[9] Miller JS, Burdick JA. Editorial: special issue on 3D printing of biomaterials. ACS Biomater Sci Eng 2016;2:1658−61.

[10] Li J, Chen M, Fan X, Zhou H. Recent advances in bioprinting techniques: approaches, applications and future prospects. J Transl Med 2016;14:271.

[11] Cui X, Dean D, Ruggeri ZM, Boland T. Cell damage evaluation of thermal inkjet printed Chinese hamster ovary cells. Biotechnol Bioeng 2010;106:963−9.

[12] Cui X, Boland T, D'Lima DD, Lotz MK. Thermal inkjet printing in tissue engineering and regenerative medicine. Recent Pat Drug Deliv Formul 2012;6:149−55.

[13] Nakamura M, Kobayashi A, Takagi F, et al. Biocompatible inkjet printing technique for designed seeding of individual living cells. Tissue Eng 2005;11:1658−66.

[14] Saunders RE, Gough JE, Derby B. Delivery of human fibroblast cells by piezoelectric drop-on-demand inkjet printing. Biomaterials 2008;29:193−203.

[15] Okamoto T, Suzuki T, Yamamoto N. Microarray fabrication with covalent attachment of DNA using bubble jet technology. Nat Biotechnol 2000;18:438−41.

[16] Hockaday LA, Kang KH, Colangelo NW, et al. Rapid 3D printing of anatomically accurate and mechanically heterogeneous aortic valve hydrogel scaffolds. Biofabrication 2012;4:035005.

[17] Geng L, Feng W, Hutmacher DW, Wong YS, Loh HT, Fuh JYH. Direct writing of chitosan scaffolds using a robotic system. Rapid Prototyping J 2005;11:90−7.

[18] Mironov V, Reis N, Derby B. Review: bioprinting: a beginning. Tissue Eng 2006;12: 631−4.

[19] Ning L, Chen X. A brief review of extrusion-based tissue scaffold bio-printing. Biotechnol J 2017;12.

[20] Chen XB. Modeling of rotary screw fluid dispensing processes. J Electron Packag 2006;129:172−8.

[21] Nishiyama Y, Nakamura M, Henmi C, et al. Development of a three-dimensional bioprinter: construction of cell supporting structures using hydrogel and state-of-the-art inkjet technology. J Biomech Eng 2009;131:035001.

[22] Wang X, Rijff BL, Khang G. A building-block approach to 3D printing a multichannel, organ-regenerative scaffold. J Tissue Eng Regener Med 2017;11:1403−11.

[23] Chien KB, Makridakis E, Shah RN. Three-dimensional printing of soy protein scaffolds for tissue regeneration. Tissue Eng C Methods 2013;19:417−26.

[24] Lim TC, Chian KS, Leong KF. Cryogenic prototyping of chitosan scaffolds with controlled micro and macro architecture and their effect on in vivo neovascularization and cellular infiltration. J Biomed Mater Res 2010;94:1303−11.

[25] Catros S, Fricain JC, Guillotin B, et al. Laser-assisted bioprinting for creating on-demand patterns of human osteoprogenitor cells and nano-hydroxyapatite. Biofabrication 2011;3:025001.

[26] Guillemot F, Souquet A, Catros S, et al. High-throughput laser printing of cells and biomaterials for tissue engineering. Acta Biomater 2010;6:2494−500.

[27] Shaw-Stewart JRH, Lippert TK, Nagel M, Nüesch FA, Wokaun A. Sequential printing by laser-induced forward transfer to fabricate a polymer light-emitting diode pixel. ACS Appl Mater Interfaces 2012;4:3535−41.

[28] Barron JA, Krizman DB, Ringeisen BR. Laser printing of single cells: statistical analysis, cell viability, and stress. Ann Biomed Eng 2005;33:121−30.

[29] Othon CM, Wu X, Anders JJ, Ringeisen BR. Single-cell printing to form three-dimensional lines of olfactory ensheathing cells. Biomed Mater 2008;3:034101.

[30] Barron JA, Wu P, Ladouceur HD, Ringeisen BR. Biological laser printing: a novel technique for creating heterogeneous 3-dimensional cell patterns. Biomed Microdevices 2004;6:139−47.

[31] Ringeisen BR, Kim H, Barron JA, et al. Laser printing of pluripotent embryonal carcinoma cells. Tissue Eng 2004;10:483−91.

[32] Guillotin B, Souquet A, Catros S, et al. Laser assisted bioprinting of engineered tissue with high cell density and microscale organization. Biomaterials 2010;31:7250−6.

[33] Murphy SV, Skardal A, Atala A. Evaluation of hydrogels for bio-printing applications. J Biomed Mater Res 2013;101A:272−84.

[34] Melchels FP, Feijen J, Grijpma DW. A review on stereolithography and its applications in biomedical engineering. Biomaterials 2010;31:6121−30.

[35] Chia HN, Wu BM. Recent advances in 3D printing of biomaterials. J Biol Eng 2015;9:4.

[36] Park JH, Jang J, Lee JS, Cho DW. Three-dimensional printing of tissue/organ analogues containing living cells. Ann Biomed Eng 2017;45:180−94.

[37] Zongjie W, Raafa A, Benjamin P, Roya S, Sanjoy G, Keekyoung K. A simple and high-resolution stereolithography-based 3D bioprinting system using visible light crosslinkable bioinks. Biofabrication 2015;7:045009.

[38] Odde DJ, Renn MJ. Laser-guided direct writing of living cells. Biotechnol Bioeng 2000;67:312−8.

[39] Zhu W, Wang M, Fu Y, Castro NJ, Fu SW, Zhang LG. Engineering a biomimetic three-dimensional nanostructured bone model for breast cancer bone metastasis study. Acta Biomater 2015;14:164−74.

[40] Gu BK, Choi DJ, Park SJ, Kim MS, Kang CM, Kim CH. 3-Dimensional bioprinting for tissue engineering applications. Biomater Res 2016;20:12.

[41] Ali M, Pages E, Ducom A, Fontaine A, Guillemot F. Controlling laser-induced jet formation for bioprinting mesenchymal stem cells with high viability and high resolution. Biofabrication 2014;6:045001.

[42] Skoog SA, Goering PL, Narayan RJ. Stereolithography in tissue engineering. J Mater Sci Mater Med 2014;25:845−56.

[43] Ji S, Guvendiren M. Recent advances in bioink design for 3D bioprinting of tissues and organs. Front Bioeng Biotechnol 2017;5:23.

[44] Gudapati H, Dey M, Ozbolat I. A comprehensive review on droplet-based bioprinting: past, present and future. Biomaterials 2016;102:20−42.

[45] Katja H, Shengmao L, Liesbeth T, Sandra Van V, Linxia G, Aleksandr O. Bioink properties before, during and after 3D bioprinting. Biofabrication 2016;8:032002.

[46] Rucker M, Laschke MW, Junker D, et al. Angiogenic and inflammatory response to biodegradable scaffolds in dorsal skinfold chambers of mice. Biomaterials 2006;27:5027−38.

[47] Fedorovich NE, Alblas J, de Wijn JR, Hennink WE, Verbout AJ, Dhert WJ. Hydrogels as extracellular matrices for skeletal tissue engineering: state-of-the-art and novel application in organ printing. Tissue Eng 2007;13:1905—25.

[48] Pan T, Song W, Cao X, Wang Y. 3D bioplotting of gelatin/alginate scaffolds for tissue engineering: influence of crosslinking degree and pore architecture on physicochemical properties. J Mater Sci Technol 2016;32:889—900.

[49] Burdick JA, Prestwich GD. Hyaluronic acid hydrogels for biomedical applications. Adv Mater 2011;23:H41—56.

[50] Highley CB, Rodell CB, Burdick JA. Direct 3D printing of shear-thinning hydrogels into self-healing hydrogels. Adv Mater 2015;27:5075—9.

[51] Kang HW, Lee SJ, Ko IK, Kengla C, Yoo JJ, Atala AA. 3D bioprinting system to produce human-scale tissue constructs with structural integrity. Nat Biotechnol 2016;34:312—9.

[52] Wust S, Muller R, Hofmann S. 3D Bioprinting of complex channels-effects of material, orientation, geometry, and cell embedding. J Biomed Mater Res 2015;103:2558—70.

[53] Hribar KC, Soman P, Warner J, Chung P, Chen S. Light-assisted direct-write of 3D functional biomaterials. Lab Chip 2014;14:268—75.

[54] Wilson Jr WC, Boland T. Cell and organ printing 1: protein and cell printers. Anat Rec A Discov Mol Cell Evol Biol 2003;272:491—6.

[55] Mironov V, Boland T, Trusk T, Forgacs G, Markwald RR. Organ printing: computer-aided jet-based 3D tissue engineering. Trends Biotechnol 2003;21:157—61.

[56] Christensen K, Xu C, Chai W, Zhang Z, Fu J, Huang Y. Freeform inkjet printing of cellular structures with bifurcations. Biotechnol Bioeng 2015;112:1047—55.

[57] Jakab K, Neagu A, Mironov V, Markwald RR, Forgacs G. Engineering biological structures of prescribed shape using self-assembling multicellular systems. Proc Natl Acad Sci U S A 2004;101:2864—9.

[58] Norotte C, Marga FS, Niklason LE, Forgacs G. Scaffold-free vascular tissue engineering using bioprinting. Biomaterials 2009;30:5910—7.

[59] Nguyen DG, Funk J, Robbins JB, et al. Bioprinted 3D primary liver tissues allow assessment of organ-level response to clinical drug induced toxicity in vitro. PLoS One 2016;11:e0158674.

[60] Pati F, Jang J, Ha DH, et al. Printing three-dimensional tissue analogues with decellularized extracellular matrix bioink. Nat Commun 2014;5:3935.

[61] Jang J, Yi H-G, Cho D-W. 3D printed tissue models: present and future. ACS Biomater Sci Eng 2016;2:1722—31.

[62] Lee H, Han W, Kim H, et al. Development of liver decellularized extracellular matrix bioink for three-dimensional cell printing-based liver tissue engineering. Biomacromolecules 2017;18:1229—37.

[63] Ker ED, Nain AS, Weiss LE, et al. Bioprinting of growth factors onto aligned submicron fibrous scaffolds for simultaneous control of cell differentiation and alignment. Biomaterials 2011;32:8097—107.

[64] Pike DB, Cai S, Pomraning KR, et al. Heparin-regulated release of growth factors in vitro and angiogenic response in vivo to implanted hyaluronan hydrogels containing VEGF and bFGF. Biomaterials 2006;27:5242—51.

[65] Zhu J, Marchant RE. Design properties of hydrogel tissue-engineering scaffolds. Expet Rev Med Dev 2011;8:607—26.

[66] Chen FM, Zhang M, Wu ZF. Toward delivery of multiple growth factors in tissue engineering. Biomaterials 2010;31:6279—308.

[67] Lee K, Silva EA, Mooney DJ. Growth factor delivery-based tissue engineering: general approaches and a review of recent developments. J R Soc Interface 2011;8: 153—70.

[68] Tasoglu S, Demirci U. Bioprinting for stem cell research. Trends Biotechnol 2013;31: 10—9.

[69] Cheung DY, Duan B, Butcher JT. Current progress in tissue engineering of heart valves: multiscale problems, multiscale solutions. Expet Opin Biol Ther 2015;15: 1155—72.

[70] Kallio M, Pihkala J, Sairanen H, Mattila I. Long-term results of the Ross procedure in a population-based follow-up. Eur J Cardiothorac Surg 2015;47:e164—70.

[71] Butcher JT, Mahler GJ, Hockaday LA. Aortic valve disease and treatment: the need for naturally engineered solutions. Adv Drug Deliv Rev 2011;63:242—68.

[72] Hinton RB, Yutzey KE. Heart valve structure and function in development and disease. Annu Rev Physiol 2011;73:29—46.

[73] Kershaw JD, Misfeld M, Sievers HH, Yacoub MH, Chester AH. Specific regional and directional contractile responses of aortic cusp tissue. J Heart Valve Dis 2004;13: 798—803.

[74] Thubrikar M, Piepgrass WC, Deck JD, Nolan SP. Stresses of natural versus prosthetic aortic valve leaflets in vivo. Ann Thorac Surg 1980;30:230—9.

[75] Leeson-Dietrich J, Boughner D, Vesely I. Porcine pulmonary and aortic valves: a comparison of their tensile viscoelastic properties at physiological strain rates. J Heart Valve Dis 1995;4:88—94.

[76] Dagum P, Green GR, Nistal FJ, et al. Deformational dynamics of the aortic root: modes and physiologic determinants. Circulation 1999;100:II54—62.

[77] Combs MD, Yutzey KE. Heart valve development: regulatory networks in development and disease. Circ Res 2009;105:408—21.

[78] Bertipaglia B, Ortolani F, Petrelli L, et al. Cell characterization of porcine aortic valve and decellularized leaflets repopulated with aortic valve interstitial cells: the VESA-LIO Project (Vitalitate Exornatum Succedaneum Aorticum Labore Ingenioso Obtenibitur). Ann Thorac Surg 2003;75:1274—82.

[79] Jana S, Tranquillo RT, Lerman A. Cells for tissue engineering of cardiac valves. J Tissue Eng Regener Med 2016;10:804—24.

[80] Benton JA, Kern HB, Anseth KS. Substrate properties influence calcification in valvular interstitial cell culture. J Heart Valve Dis 2008;17:689—99.

[81] Jana S, Tefft BJ, Spoon DB, Simari RD. Scaffolds for tissue engineering of cardiac valves. Acta Biomater 2014;10:2877—93.

[82] Weber B, Emmert MY, Schoenauer R, Brokopp C, Baumgartner L, Hoerstrup SP. Tissue engineering on matrix: future of autologous tissue replacement. Semin Immunopathol 2011;33:307—15.

[83] Ravi S, Chaikof EL. Biomaterials for vascular tissue engineering. Regen Med 2010;5: 107—20.

[84] Charitos EI, Sievers H-H. Anatomy of the aortic root: implications for valve-sparing surgery. Ann Cardiothorac Surg 2013;2:53—6.

[85] Sahasakul Y, Edwards WD, Naessens JM, Tajik AJ. Age-related changes in aortic and mitral valve thickness: implications for two-dimensional echocardiography based on an autopsy study of 200 normal human hearts. Am J Cardiol 1988;62:424−30.

[86] Duan B, Hockaday LA, Kang KH, Butcher JT. 3D bioprinting of heterogeneous aortic valve conduits with alginate/gelatin hydrogels. J Biomed Mater Res 2013;101:1255−64.

[87] Novosel EC, Kleinhans C, Kluger PJ. Vascularization is the key challenge in tissue engineering. Adv Drug Deliv Rev 2011;63:300−11.

[88] Kinstlinger IS, Miller JS. 3D-printed fluidic networks as vasculature for engineered tissue. Lab Chip 2016;16:2025−43.

[89] Ko HC, Milthorpe BK, McFarland CD. Engineering thick tissues—the vascularisation problem. Eur Cell Mater 2007;14:1−18. discussion 18−19.

[90] Kaully T, Kaufman-Francis K, Lesman A, Levenberg S. Vascularization—the conduit to viable engineered tissues. Tissue Eng Part B Rev 2009;15:159−69.

[91] Datta P, Ayan B, Ozbolat IT. Bioprinting for vascular and vascularized tissue biofabrication. Acta Biomater 2017;51:1−20.

[92] Neufurth M, Wang X, Tolba E, et al. Modular small diameter vascular grafts with bioactive functionalities. PLoS One 2015;10:e0133632.

[93] Pashneh-Tala S, MacNeil S, Claeyssens F. The tissue-engineered vascular graft-past, present, and future. Tissue Eng Part B Rev 2016 Feb 1;22(1):68−100.

[94] Thottappillil N, Nair PD. Scaffolds in vascular regeneration: current status. Vasc Health Risk Manag 2015;11:79−91.

[95] Mulligan-Kehoe MJ, Simons M. Vasa vasorum in normal and diseased arteries. Circulation 2014;129:2557−66.

[96] Nemeno-Guanzon JG, Lee S, Berg JR, et al. Trends in tissue engineering for blood vessels. J Biomed Biotechnol 2012;2012:956345.

[97] Elliott MB, Gerecht S. Three-dimensional culture of small-diameter vascular grafts. J Mater Chem B 2016;4:3443−53.

[98] Jakab K, Norotte C, Marga F, Murphy K, Vunjak-Novakovic G, Forgacs G. Tissue engineering by self-assembly and bio-printing of living cells. Biofabrication 2010;2:022001.

[99] Hinton TJ, Jallerat Q, Palchesko RN, et al. Three-dimensional printing of complex biological structures by freeform reversible embedding of suspended hydrogels. Sci Adv 2015;1:e1500758.

[100] Bertassoni LE, Cecconi M, Manoharan V, et al. Hydrogel bioprinted microchannel networks for vascularization of tissue engineering constructs. Lab Chip 2014;14:2202−11.

[101] Lee VK, Kim DY, Ngo H, et al. Creating perfused functional vascular channels using 3D bio-printing technology. Biomaterials 2014;35:8092−102.

[102] Panwar A, Tan LP. Current status of bioinks for micro-extrusion-based 3D bioprinting. Molecules 2016:21.

[103] Shengjie L, Zhuo X, Xiaohong W, Yongnian Y, Haixia L, Renji Z. Direct fabrication of a hybrid cell/hydrogel construct by a double-nozzle assembling technology. J Bioact Compat Polym 2009;24:249−65.

[104] Skardal A, Zhang J, Prestwich GD. Bioprinting vessel-like constructs using hyaluronan hydrogels crosslinked with tetrahedral polyethylene glycol tetracrylates. Biomaterials 2010;31:6173−81.

[105] Yu Y, Zhang Y, Martin JA, Ozbolat IT. Evaluation of cell viability and functionality in vessel-like bioprintable cell-laden tubular channels. J Biomech Eng 2013;135: 91011.

[106] Gao Q, He Y, Fu JZ, Liu A, Ma L. Coaxial nozzle-assisted 3D bioprinting with built-in microchannels for nutrients delivery. Biomaterials 2015;61:203—15.

[107] Zhang Y, Yu Y, Chen H, Ozbolat IT. Characterization of printable cellular micro-fluidic channels for tissue engineering. Biofabrication 2013;5:025004.

[108] Zhang Y, Yu Y, Akkouch A, Dababneh A, Dolati F, Ozbolat IT. In vitro study of directly bioprinted perfusable vasculature conduits. Biomater Sci 2015;3:134—43.

[109] Kesari P, Xu T, Boland T. Layer-by-layer printing of cells and its application to tissue engineering. MRS Proc 2011:845.

[110] Cui X, Boland T. Human microvasculature fabrication using thermal inkjet printing technology. Biomaterials 2009;30:6221—7.

[111] Wu PK, Ringeisen BR. Development of human umbilical vein endothelial cell (HUVEC) and human umbilical vein smooth muscle cell (HUVSMC) branch/stem structures on hydrogel layers via biological laser printing (BioLP). Biofabrication 2010;2:014111.

[112] Pirlo RK, Wu P, Liu J, Ringeisen B. PLGA/hydrogel biopapers as a stackable substrate for printing HUVEC networks via BioLP. Biotechnol Bioeng 2012;109:262—73.

[113] Xiong R, Zhang Z, Chai W, Huang Y, Chrisey DB. Freeform drop-on-demand laser printing of 3D alginate and cellular constructs. Biofabrication 2015;7:045011.

[114] Hribar KC, Meggs K, Liu J, Zhu W, Qu X, Chen S. Three-dimensional direct cell patterning in collagen hydrogels with near-infrared femtosecond laser. Sci Rep 2015;5:17203.

[115] Dahl SLM, Rhim C, Song YC, Niklason LE. Mechanical properties and compositions of tissue engineered and native arteries. Ann Biomed Eng 2007;35:348—55.

[116] Kelm JM, Djonov V, Ittner LM, et al. Design of custom-shaped vascularized tissues using microtissue spheroids as minimal building units. Tissue Eng 2006;12:2151—60.

[117] Cui H, Nowicki M, Fisher JP, Zhang LG. 3D bioprinting for organ regeneration. Adv Healthc Mater 2017:6.

[118] Yeh RW, Sidney S, Chandra M, Sorel M, Selby JV, Go AS. Population trends in the incidence and outcomes of acute myocardial infarction. N Engl J Med 2010;362: 2155—65.

[119] Duan B. State-of-the-Art review of 3D bioprinting for cardiovascular tissue engineering. Ann Biomed Eng 2017;45:195—209.

[120] Murry CE, Wiseman RW, Schwartz SM, Hauschka SD. Skeletal myoblast transplantation for repair of myocardial necrosis. J Clin Invest 1996;98:2512—23.

[121] Udelson JE, Patten RD, Konstam MA. New concepts in post-infarction ventricular remodeling. Rev Cardiovasc Med 2003;4(Suppl. 3):S3—12.

[122] Leor J, Aboulafia-Etzion S, Dar A, et al. Bioengineered cardiac grafts: a new approach to repair the infarcted myocardium? Circulation 2000;102:III56—61.

[123] Gaetani R, Barile L, Forte E, et al. New perspectives to repair a broken heart. Cardiovasc Hematol Agents Med Chem 2009;7:91—107.

[124] Gaetani R, Rizzitelli G, Chimenti I, et al. Cardiospheres and tissue engineering for myocardial regeneration: potential for clinical application. J Cell Mol Med 2010;14: 1071—7.

[125] Gaetani R, Doevendans PA, Metz CH, et al. Cardiac tissue engineering using tissue printing technology and human cardiac progenitor cells. Biomaterials 2012;33: 1782—90.

[126] Gaetani R, Feyen DA, Verhage V, et al. Epicardial application of cardiac progenitor cells in a 3D-printed gelatin/hyaluronic acid patch preserves cardiac function after myocardial infarction. Biomaterials 2015;61:339—48.

[127] Gaebel R, Ma N, Liu J, et al. Patterning human stem cells and endothelial cells with laser printing for cardiac regeneration. Biomaterials 2011;32:9218—30.

[128] Rodell CB, MacArthur JW, Dorsey SM, et al. Shear-thinning supramolecular hydrogels with secondary autonomous covalent crosslinking to modulate viscoelastic properties in vivo. Adv Funct Mater 2015;25:636—44.

Multimaterial Cardiovascular Printing

10

Bobak Mosadegh[1,2], Ahmed Amro[3], Yazan Numan[3]

Department of Radiology, Weill Cornell Medicine, New York, NY, United States[1]; Dalio Institute of Cardiovascular Imaging, NewYork-Presbyterian Hospital, New York, NY, United States[2]; Joan C. Edwards School of Medicine, Marshall University, Huntington, West Virginia, United States[3]

INTRODUCTION

Tissues are hierarchical structures comprising many different materials that allow them to adapt to various internal and external stressors. As a direct result, the location and amount of constituent tissue varies as a result of the presence of health or a particular disease condition. Ultimate organ function depends on tissue composition and properties [1]. The cardiovascular system can be broadly divided into three components: myocardium, cardiac valves, and blood vessels. In the context of multimaterial printing, the materials being used depend on the intended application. 3D printed synthetic models are useful for visualizing the complex anatomy of the heart, the relative positioning of different structures or tissue types, or for how cardiac devices and tools will interface with the heart during a procedure. These models can be useful for both educational and clinical applications [2]. 3D printed biological models have the potential to be useful for understanding how cells interact in more physiologic environments, and as therapeutic implants for tissue engineering applications. For both synthetic and biological 3D printed models, the ability to 3D print multiple materials is very important in order to recapitulate the complexities of the heart. 3D printing has the advantage of fabricating complex structures using a layered approach [3]. For cardiovascular applications, this technology can be used for a variety of purposes, such as 3D models to visualize anatomy, surgical planning for device deployment, and direct fabrication of tissues [4].

Two types of 3D printing technologies enable robust fabrication of objects comprising multiple materials for cardiovascular applications: (1) material extrusion, which uses nozzles on a gantry to lay down filaments of a material that bonds to the surrounding material either by cooling, UV-activation, or evaporation, and (2) material jetting, which expels an array of material droplets and cures them using UV light. Material extrusion is primarily used to fabricate plastic models for visualizing cardiac anatomy for preprocedural planning. By integrating multiple nozzles in a fused deposition modeling printer, materials of different stiffness and color can better visualize the heterogeneous nature of cardiac structures and tissue properties. For

3D Printing Applications in Cardiovascular Medicine. https://doi.org/10.1016/B978-0-12-803917-5.00010-9

applications involving bioprinting of cells, multiple nozzles can be integrated to deposit various types of hydrogels and cells to recapitulate the diversity of cardiac tissue. For material jetting, the PolyJet technology by Stratasys is the most advanced, provided by their line of Objet Connex printers that provide high resolution (e.g., down to 16 microns) and the ability to perform digital printing, which enables a gradient of material stiffness and extensibility to be fabricated within a given model. The most popular material mixtures are between the Vero (rigid) and Tango (soft) materials. Furthermore, Stratasys recently launched a voxel printing technology using GrabCAD Voxel Print on the J750 printer. This technology enables each voxel to be specified with a unique material property, enabling unprecedented variation in stiffness, extensibility, and transparency.

A key requirement for printing multiple materials is knowledge of the geometry and spatial material composition of the desired tissue [5]. Although specific details are still under investigation, much of this information is well characterized for a generalized or ideal tissue [6−8], but to acquire this knowledge for a particular patient, medical imaging is used to noninvasively map the geometry and properties of the tissue using a myriad of techniques (e.g., CT, MRI, echocardiography, cardiac-PET, etc.). Fig. 10.1 summarizes the overall procedure for generating a 3D printable file from medical images [9]. For multimaterial printing, these files must be segmented in a manner that distinguishes the different tissues or structures of interest. These standard tessellation language (STL) files can then be merged to generate the overall desired structure, but with each STL file set to print a different material. Software is being developed that helps facilitate the programming of these complex files [10]. Currently, the types of materials available in multimaterial printers is limited, ranging in tensile strength from ∼0.8 to 45 MPa [11] or ∼3 to 200 kPa [12,13] for Stratasys PolyJet-based 3D printers or hydrogel-based extrusion 3D printers, respectively. The use of multimaterial 3D printing is described below, categorized by its use for specific cardiovascular applications. Table 10.1 lists recent publications that utilize multiple materials for creating cardiovascular models and bioprinted tissues.

MULTIMATERIAL PRINTING FOR EDUCATION AND COMMUNICATION

One potential application of 3D printed models is as an educational tool. Currently 3D printed models used for educational purposes have been primarily single material models [14−18]. For example, 3D printing of a thoracic aortic aneurysm model from a patient's CT scan has been frequently used to educate cardiovascular surgery residents prior to the actual surgical procedure [19]. Such a model has indeed helped residents understand the various small anatomical details that would otherwise be obscure or unnoticeable, while reviewing a CT scan. In addition, these models have helped avoid surgical complications as a result of in-depth understanding of

Procedural Steps

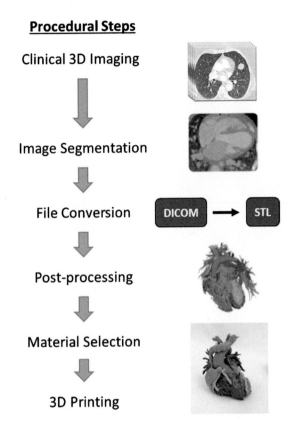

Clinical 3D Imaging

Image Segmentation

File Conversion

Post-processing

Material Selection

3D Printing

FIGURE 10.1

Step-by-step approach to multimaterial 3D printing. The initial approach involves determination of the appropriate images to be utilized, followed by segmentation of each desired feature, and conversion of the file into an STL format with each segmentation in the same coordinate space. Postprocessing steps that ensure the .STL file is watertight, smoothed, and trimmed. In the 3D printing software, materials are selected for each .STL file as desired and set to print. *DICOM*, digital imaging communications in medicine; *STL*, standard tessellation language.

the anatomy of a particular aneurysm. Multimaterial printing, however, is crucial for highlighting different regions of interest by printing different colors within a model (Fig. 10.2). Although most educational models are comprised of a single type of material (in terms of stiffness and/or extensibility), as multimaterial printing technology and available materials improve, 3D printed models will better capture the differences in material properties for various tissues. This fact is supported by a study that evaluated the usefulness of 3D printed cardiac models as an educational

Table 10.1 References of 3D Printed Multimaterial Models and Bioprinted Tissues

Application	First Author (References)	Description	Type of 3D Printer	Type of Materials
Surgical planning	Giannopoulos et al. [61]	Cardiothoracic anatomic models	N/A	N/A
	Yang et al. [62]	Hypertrophic cardiomyopathy model	Material jetting; PolyJet	Tango series
	Mahmood et al. [25]	Mitral annulus and leaflets	Material jetting; PolyJet	Rigid and soft
	Vukicevic et al. [26,27,63]	Mitral valve	Material jetting; PolyJet	Multiple vero and tango materials
	Little et al. [23]	Mitral valve	Material jetting; PolyJet	Multiple vero and tango materials
	Izzo et al. [24]	Mitral valve	Material jetting; PolyJet	Tango plus; DM_9770
	Maragiannis et al. [64,65]	Replicating severe aortic stenosis	Material jetting; PolyJet	Multiple vero and tango materials
Cardiac valves	Nakayama et al. [41]	Implanted biosynthetic aortic valve in goats	Material jetting; 3D systems	Trileaflet valve conduit with a sinus of valsalva (biovalve type VII)
	Hockaday et al. [39]	Photo cross-linking of 3D printed hydrogel can accurately fabricate heterogeneous aortic valve	Extrusion printer	Poly-ethylene glycol-diacrylate hydrogels & alginate & porcine aortic valve interstitial cells
Cardiac tissue	Izadifar et al. [58]	UV-assisted 3D printing of cardiac tissue	Extrusion printer	Sodium alginate and human coronary artery endothelial cells
	Jang et al. [66]	Bioprinted decellularized matrix	Extrusion printer	Stem cell-laden hydrogels
	Lind et al. [60]	In vitro testing of actuation of cardiac tissue for drug testing	Extrusion printer	Dextran, polyurethane, carbon black, polyamide, silicone, gold

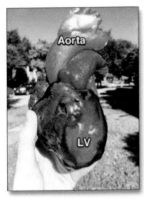 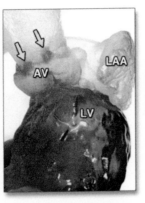 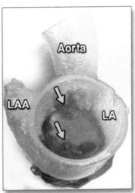

| **3D Printed Multi-color Transparent Model** | **3D Printed Multi-material Left Heart Model** | **Mitral Annulus "Calcium"** |

FIGURE 10.2

Multimaterial and multicolored patient-specific 3D printed heart for educational purposes and communication with patients. Different colors are used to visualize blood flow within the heart (right panel), and *yellow arrows* indicate regions 3D printed in pink to replicate calcium depositions within the AV (center panel) and mitral valve (right panel). *AV*, aortic valve; *LA*, left atrium; *LAA*, left atrial appendage; *LV*, left ventricle.

Reproduced with permission from Vukicevic M, et al. Cardiac 3D printing and its future directions. JACC Cardiovasc Imaging 2017;10:171–84.

tool for nurses, which found that nurses had requested models with additional colors in order to better visualize and understand the anatomy [15]. A randomized trial compared two groups that were given either a traditional cardiac plastic model or a 3D printed heart model. Even though the heart model was constructed of a single material, there was no difference in the ability of the subjects to understand the anatomy of the heart [17]. These findings suggest that 3D printed cardiac models are useful for educational purposes, and are likely to improve as models become more sophisticated.

There is a practical question of whether the advantages afforded by 3D printing are particularly useful for such an application, over traditional manufacturing methods that generate models in bulk. 3D printing provides the advantage of making unique models in a cost-effective manner, and in the case of cardiovascular models, this capability is most useful for fabricating patient-specific models. Educational models, however, don't necessarily need to be patient-specific since they are used for teaching purposes. On the other hand, as multimaterial printing improves, models that integrate complex shapes and material properties that recapitulate tissue properties may become more cost-effective to produce using 3D printing technology.

MULTIMATERIAL PRINTING FOR SURGICAL PLANNING

3D printing has improved surgical planning by providing a physical model to visualize the complex cardiac structural abnormalities. The availability of these 3D models before surgical intervention allows clear demarcation of normal from abnormal tissue, thus assisting with reducing the number of unanticipated complications. Using 3D volumetric image data generated from CT, cardiac magnetic resonance, or 3D echocardiography, a patient-specific cardiac model can be fabricated as an intuitive method for perceiving cardiac anatomy and pathology with full-depth awareness and relative scaling to cardiac devices. Falk and colleagues have published a small case series on using 3D printing for cardiac surgery planning, specifically for repair of ventricular aneurysms (which commonly occurs after an anterior wall myocardial infarction). This study further examined the utility of 3D printing for resection of cardiac tumors, and found an enhanced ability to visualize tumor margins leading to adequate resection [20]. Further, Valverde and colleagues reported on the European experience and reported that 3D printing was found to be extremely helpful in planning surgery for eight patients with complex congenital heart disease [21]. Similarly, Jabbari et al. showed how 3D printed models can be used to assist in the resection of cardiac tumors [22]. These models, however, only used a single type of material or different colors within the model and did not utilize materials with varying mechanical properties to outline the heterogeneous nature of cardiac tissue.

An example of a multimaterial model was shown by Little et al., who created a patient-specific mitral valve model of a patient with severe mitral regurgitation on echocardiography, which showed leaflet malcoaptation and a perforation in the posterior mitral leaflet. The model was fabricated using CT images, where the heart tissue, leaflets, and calcifications were individually modeled using different materials [23]. This multimaterial model allowed for the appropriate sizing of an Amplatzer Duct Occluder II before implantation into the patient, along with the placement of a MitraClip (Fig. 10.3). Similarly, Izzo et al. produced a patient-specific mitral valve model from CT scans, modified the anatomy with engineering ports to be integrated into a flow loop, and then performed a mock procedure, where valve cages were implanted and imaged using fluoroscopy [24]. In such an instance, five different models and support materials were printed: (1) a soft Tango + for the vascular anatomy, (2) a rigid FLX-MT-S85DM for the support structures, (3) a rigid DM_9770 for the calcifications, (4) insoluble support material, and (5) a soluble support material.

Another example of using a multimaterial printer for surgical planning was demonstrated by Mahmood et al., who created a mitral valve model based on an imaging dataset acquired directly from 3D transesophageal echocardiography (TEE) [25]. The high temporal resolution of TEE, coupled with gated imaging, enabled precise interrogation of mitral valve leaflet geometries, along with the mitral valve annulus. The authors were able to fabricate models of normal, ischemic, and

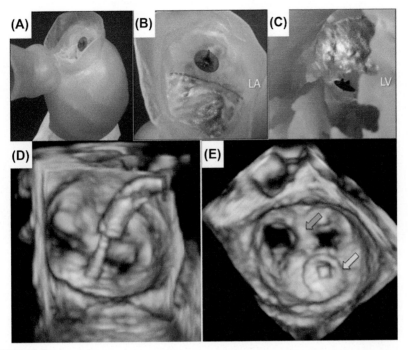

FIGURE 10.3

Multimaterial mitral valve model. (A—C) Different views of the model containing leaflets and calcification for the surgical planning of an Amplatzer occluder deployment to fill a posterior leaflet perforation. (D—E) 3D echocardiographic images after implantation into the patient. *Green arrow* points to the MitraClip and *yellow arrow* points to the occluder. *LA*, left atrium; *LV*, left ventricle.

Reproduced with permission from Little SH, et al. 3D printed modeling for patient-specific mitral valve intervention: repair with a clip and a plug. JACC Cardiovasc Interv 2016;9:973—5.

myxomatous mitral valves within 90 min. However, the thickness of the anatomic structures were not accurate due to limitations of the printer and material properties. However, the authors did show that mitral valve models could be made using all soft materials or a combination of a rigid annulus and soft leaflets.

Recently, more sophisticated examples of multimaterial 3D printed models for surgical planning was shown by Vukicevic et al. [26]. The major highlight of their work was to use multiple Tango materials to generate a synthetic leaflet composite with similar slope of the stress-strain toe regions to a native mitral valve tissue. Related work, as seen in Fig. 10.4, shows how a mitral valve can be 3D printed using multiple materials to distinguish the properties of its annulus, chordae, and papillary muscles in order to test the benchtop implantation of a MitraClip device [27].

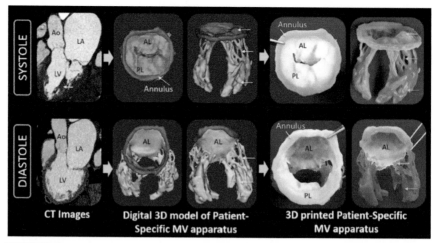

FIGURE 10.4

MV apparatus reconstruction from CT imaging. Systolic (upper left) and diastolic CT images (lower left) are used for the digital reconstruction of mitral valve apparatus in systole and diastole (middle panel), including the mitral annulus, anterior and posterior leaflet (AL, PL), chordae and papillary muscles (Pap). Eventually the model of mitral valve apparatus was 3D printed of multiple materials, with harder annulus (*pink*), flexible leaflets (*translucent white*), flexible chordae (*blue*) and more rigid papillary muscles (*blue*) (right panels). *AL*, anterior leaflet; *Ao*, aorta; *LA*, left atrium; *LV*, left ventricle; *MV*, mitral valve, *PL*, posterior leaflet.

Reproduced with permission from Vukicevic M, et al. 3D printed patient-specific multi-material modeling of mitral valve apparatus. J Am Coll Cardiol 2017;69:1128.

MULTIMATERIAL PRINTING OF VALVES

Embryonically, the heart valves are formed by the endocardial cushions, which are mesenchymal in origin. The two atrioventricular (AV) valves are the tricuspid and mitral valves, while the two ventriculovascular track (VV) valves are the aortic and pulmonic valves. All four valves have two main components: cellular and extracellular matrix (ECM).

Cellular components: The most frequent cells found in the valve tissue are the valvular interstitial cells (VICs). These cells are responsible for the maintenance and regeneration of cellular and matrix components of the valve. Morphologically, these cells are elongated and slender in shape, are involved in matrix production, and possess a prominent rough endoplasmic reticulum along with Golgi apparatus. VICs can also become contractile, due to activation by TGF-β, and are involved in maintaining hemodynamic forces across the valve. The second prominent cell type are the valvular endocardial cells (VECs); these cells are located across the entire surface of the valve. These cells are crucial to valve physiology as they negate thrombosis. Cardiac, smooth muscle, and neural cells are other minor cellular components [28].

ECM component: There are three morphologically and functionally distinct layers in the valve leaflets; lamina fibrosa, lamina spongiosa, and lamina ventricularis in the VV valves. The AV valves are similar to the VV valves, except that the ventricularis layer is replaced by a thin atrialis layer. These three major layers are formed by varying proportions of four major macromolecules, where each one confers a unique physiologic ability. The most abundant macromolecule is collagen. Collagen accounts for more than 60% of the weight of the valve, and provides the mechanical stability and strength of the valve. The most abundant collagen types are type I and III, while type V is found only to a minor degree. These collagen macromolecules are surrounded by elastin, which is the second most abundant component and provides interconnections between collagen fibers. The third component is glycosaminoglycan (GAG), which functions as the gel-like ground substance for the matrix. The major GaGs are hyaluronic acid, dermatan sulfate, and chondroitin sulfate. At the attachment of the leaflet to the mitral annulus, the three matrix layers of the leaflet morph into a single fibrous layer that constitutes the major component of the annulus. The differential spreading and continuation of the fibrosa layer to the annulus provides the tensile strength needed for valve anchoring [29].

Other structures related to the valve are the papillary muscles and chordae tendineae. Chordae tendineae are composed of several types of ECM proteins, such as collagen type I, III, and fibronectin in the spongiosa layer, and collagen type III resides in the fibrosa layer [30]. Chordae tendineae, when under tension, have a secant modulus of ~ 100 MPa [31]. The primary cell types in chordae tendineae are endothelial and fibroblast cells. Papillary muscles, which are present only in VV valves, are cylindrical extensions of the myocardial tissue that connect to the chordae tendineae, such that leaflets do not fold backwards during ventricular contractions [32].

The Butcher lab is leading the effort for 3D printing of heart valves using cells and ECM scaffolding [33–40]. The primary mechanism for printing is extrusion of photocurable hydrogels that contain cells of a desired type, with different mixtures in separate nozzles. Fig. 10.5 shows a schematic of such a system, along with images of 3D printed valves and fluorescent images of the embedded cells. Currently these examples demonstrate the ability to 3D print structures that mimic the anatomic dimensions of the valve, but do not yet recapitulate the material properties of these tissues and, thus, are not able to be used functionally as a valve.

An alternative strategy to directly printing cell-laden hydrogels to fabricate a valve is to use 3D printed parts as molds to grow a valve subcutaneously into a desired shape in vitro [41]. In this work, 3D printed parts were assembled such that their negative volume created the desired tissue shape for a valve. These molds were implanted subcutaneously into goats for 2 months, where tissue grew into the free spaces of the assembled 3D printed molds (Fig. 10.6). Once explanted, these "biovalves" were able to be reimplanted into the aortic valve of the same animal. This approach demonstrates a unique view of how 3D printed materials can be integrated with biological tissue in vivo to create functional cardiac devices for therapy.

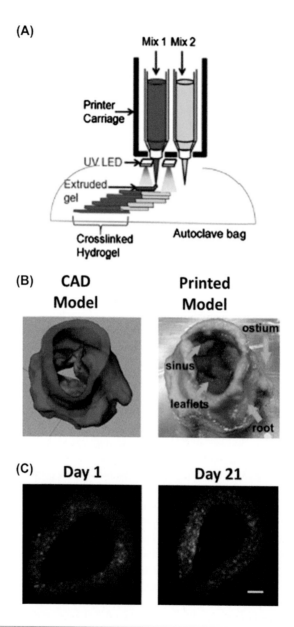

FIGURE 10.5

Bioprinted heart valve. (A) Schematic of bioprinter, consisting of two extrusion nozzles that have integrated UV light sources to cross-link the hydrogel. (B) Images of the CAD model and bioprinted valve, whose geometry was based on a micro CT scan. *Red* represents leaflets and *blue* represents the valve root. (C) Live (*fluorescent green*) and dead (*red*) images of the valve root at the annulus. Scale bar: 2 mm. *CAD,* computed aided-design; *UV,* ultraviolet.

Reproduced with permission from Hockaday LA, et al. Rapid 3D printing of anatomically accurate and mechanically heterogeneous aortic valve hydrogel scaffolds. Biofabrication 2012;4:035005.

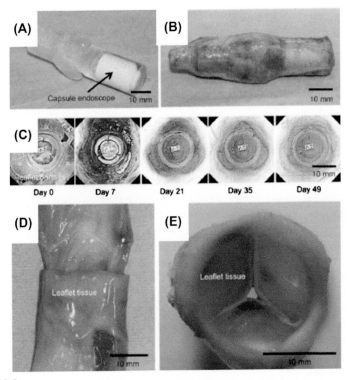

FIGURE 10.6

Regenerated heart valve from 3D printed mold, termed Biovalve. A) Image of 3D printed mold, containing capsule for endoscope to visualize tissue growth. B) Image of 3D printed mold after ~2 months of tissue growth. C) Timelapse images of tissue growth as viewed by the endoscope. D) Image of the luminal surface of the Biovalve. E) Image of the sinus Valsalva of the biovalve.

Reproduced from Nakayama Y, et al. In-body tissue-engineered aortic valve (biovalve type VII) architecture based on 3D printer molding. J Biomed Mater Res B Appl Biomater 2015;103:1−11.

MULTIMATERIAL PRINTING OF CORONARIES

Coronary artery bypass grafting (CABG) operations are performed nearly 400,000 times annually in the US. Usually native vessels from the patient are excised as grafts to be implanted, in order to bypass a blockage in the patient's coronary arteries. Candidate vessels include the internal thoracic arteries, radial arteries, or saphenous veins. Although CABG can improve survival of patients with coronary artery disease, about a third of patients do not have eligible vessels to serve as the graft. There are many drawbacks associated with CABG, even if any of these vessels are available to harvest, including poor long-term patency, accidental damage to the graft during harvest, and postsurgery donor site complications [42]. Therefore, the ability to fabricate artificial grafts can address this important clinical need. The ideal vascular graft should be biocompatible, have similar compliance and density to native vessels, have high durability, and be antithrombogenic [43,44]. Overall, arteries have a stiffness on the order of ~100 kPa to 10 MPa, depending on its disease states (i.e., plaque buildup and calcification) [45].

The coronary arterial wall is composed of three layers, which is the case in all other arterial walls. These layers are the (1) tunica intima, (2) tunica media, and (3) tunica adventitia:

1. The tunica intima is the innermost and thinnest layer, which is composed of a single layer of endothelial cells and a subendothelial layer of connective tissue, called the internal elastic lamina. This elastic lamina layer is composed of interrupted layers of elastin, which separates the intima from the tunica media.
2. The tunica media, the middle and thickest layer, is composed predominantly of circular smooth muscle fibers and connective tissue (collagen, elastic fibers, and proteoglycans). The elastin is arranged in fenestrated sheets, and in between these sheets are thin layers of proteoglycan-rich ECM, collagen fibers, and smooth muscle fibers. Tunic media (due to its components of elastin, collagen, and smooth muscle cells (SMCs)) account for the majority of mechanical properties of the arterial wall. The role of this layer is to provide structural support, elasticity, vasoreactivity (i.e., vary diameter of vessel lumen to accommodate changes in blood flow). This layer is separated from the outermost layer (tunica adventitia) by a dense elastic membrane called the external elastic lamina.
3. The tunica adventitia is the outermost connective tissue layer that surrounds the vessel. This layer is composed of fibroblasts, mast cells, collagen, and elastic fibers. The primary purpose of this layer is to reinforce the radial strength of the vessel to sustain high pressures [46].

Early attempts to fabricate low-flow, small-diameter coronary grafts used expanded polytetrafluoroethylene (ePTFE aka Gortex) and woven polyethylene terephthalate (PET aka Dacron). However, grafts made from these synthetic materials resulted in poor patient outcomes due to thrombus formation. These poor results are primarily contributed to two factors: (1) undesirable hemodynamics due to kinking of the graft during implantation, and (2) improper mechanical characteristics due to compliance mismatch between the graft and tissue [44].

Currently, there are two methods for tissue engineering vascular grafts, which can be categorized into cell-only methods or those that utilize a scaffold. The cell-only methods use cultured monolayers of cells that are then rolled around a mandrel to form the sheets into a tube-like structure. By using different types of cells in the monolayers, the different layers of the coronary artery can be recapitulated. The use of a scaffold provides mechanical strength and a structure to seed cells into a larger form factor. Scaffolds can be made from materials that are either synthetic, natural ECM proteins, or ECM that has been decellularized from tissue [47]. Despite the advancement of these methods over decades, there are still no commercially available vascular grafts made from synthetic or tissue engineering approaches.

3D printing is a tool with the potential to build sophisticated structures, such as coronary arteries, by enabling the layering of different cell types to recapitulate the layers of an artery, along with the internal vasculature to sustain robust and thick medial layers. Furthermore, the ability to fabricate vessels customized for an individual can aide in the implantation of the graft, optimizing for hemodynamics and minimizing thrombosis [48]. Current research in bioprinting focuses on the in-vitro generation of vascular models in order to study the process of angiogenesis and tissue engineer vascularized thick tissue for implantation [49,50]. There is currently no method of directly bioprinting a full artificial coronary artery, but as technology advances emerge (i.e., higher resolution printing of multiple cell types and ECM layering) bioprinting the three layers should become feasible [51].

MULTIMATERIAL PRINTING OF CARDIAC TISSUE

Myocardial Infarction is the leading cause of death for the adult population in the world. The basic mechanism involves the occlusion of one of the major epicardial blood vessels that supply the myocardium with oxygenated blood. As a result, the affected area becomes ischemic, with prolonged ischemia leading to tissue death and necrosis. Subsequently, myocardial tissue is replaced by noncontractile scar tissue. One approach to repair this affected area is by direct implantation of progenitor stem cells, a treatment strategy called cellular therapy [52]. However, this technique has not been able to provide sufficient recovery of functional tissue, primarily due to inadequate cell viability upon delivery, and an inability to reverse the remodeling of the stiffened tissue. To improve outcomes, tissue engineering approaches have been focused on integrating vasculature into tissue-like constructs (e.g., cardiac patches) to improve delivery of oxygen and nutrients to the implanted cells via accelerated angiogenesis (i.e., growth of blood vessels from the host tissue and implant) [53].

Cardiac tissue is composed of cardiomyocytes, fibroblasts, and endothelial cells, along with a fibrous network of carbohydrates and proteins that comprise the ECM. The ECM of cardiac tissue is a complex network of structural (e.g., collagens, elastins) and nonstructural proteins (e.g., thrombospondin, tenascin, osteopontin, and periostin), which together form an active and dynamic environment that plays a very important role in providing physical scaffolding for the cellular components, and in facilitating and regulating events at the cellular level [54]. Overall, cardiac tissue is soft, with a modulus of E $\sim 0.02-0.5$ MPa, but can exert forces effectively due to the alignment of the cardiomyocytes, giving rise to an organized actuator [45].

3D printing has the potential to advance current tissue engineering approaches by enabling several more sophisticated abilities: precise formation of geometric patterns for scaffolding, facilitating more uniform dispersion of cell types and ECM within the tissue, patterning of multiple cell types and ECM to yield desired alignment, and integration of synthetic materials to provide enhanced functionality of the tissue. These methods of bioprinting have yielded cardiac patches that can undergo rhythmic contractions, recapitulating the dynamic nature of cardiac tissue [55–59]. Jang et al. have demonstrated how multiple stem cells (e.g., cardiac progenitor cells and mesenchymal stem cells) can be printed into meshed networks in order to prevascularize cardiac patches before implantation (Fig. 10.7).

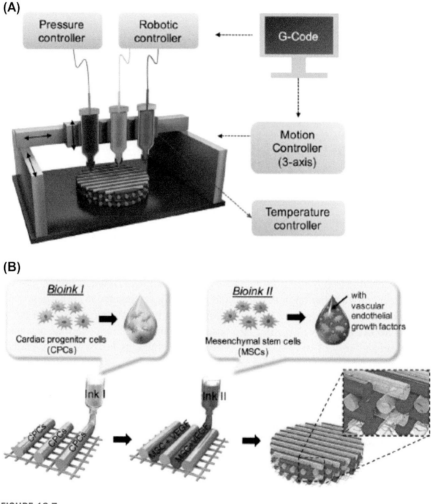

FIGURE 10.7

Schematic of prevascularized stem cell patch. (A) Illustration of 3D cell printing system, and (B) Illustration of prevascularized stem cell patch including multiple cell-laden bioinks and supporting PCL polymer.

Reproduced with permission from Jang J, et al. 3D printed complex tissue construct using stem cell-laden decellularized extracellular matrix bioinks for cardiac repair. Biomaterials 2017;112:264–74.

The advantage of printing multiple materials for cardiac patches is also being investigated to improve mechanical robustness and conductivity. Izadifar and colleagues demonstrated the integration of carbon nanotubes (CNT) into bioprinted cardiac patches that consisted of a UV-curable methacrylated collagen containing human coronary artery endothelial cells [58]. Multimaterial printing allowed distinct networks of cell-laden hydrogels to be printed along with networks of CNT-laden hydrogels. This hybrid patch was found to have an elastic modulus nearly four times greater than hydrogel alone. However, such technology is not ready for clinical use, but currently serves as a method to perform mechanistic studies, with the goal of understanding how to optimize such cell-based therapies.

A future application of multimaterial 3D printing is to build microdevices that recapitulate the microenvironment of tissue for drug testing purposes. The hope of this technology is to replace animal testing and serve as a platform that better predicts outcomes in clinical trials since drug-response to human cells can be conducted. An example of 3D printing being used for a cardiac application has been demonstrated by Lind et al. [60], whose work used an extrusion-based 3D printer to fabricate a device that senses the contraction of cardiomyocytes. This device works by using a thin cantilever that the cardiomyocytes can bend upon contraction. The degree of bending can be measured electronically via an integrated strain gauge, which was also 3D printed. To fabricate this complex device, six different materials were used, each allowing for a specific function: (1) dilate dextran as a release layer for the cantilever, (2) dilute thermoplastic polyurethane (TPU) as the cantilever, (3) mixture of carbon black and TPU as a strain gauge wire, (4) shear-thinning soft PDMS (polydimethylsiloxane) as a feature to guide tissue orientation, (5) shear-thinning mixture of gold and polyamide as electrical contact pads, (6) shear thinning PDMS and insulation (Fig. 10.8). This work serves as an example for how 3D printing multiple materials can enable multifunctional devices that can inspire the fabrication of complex tissues that will advance the field of tissue engineering.

FUTURE PERSPECTIVE OF MULTIMATERIAL PRINTING FOR CARDIOVASCULAR DEVICES

As seen above, there is a significant gap between the complexity of cardiac tissue and the ability for 3D printing technology to recapitulate its complex tissue mechanics, whether being used for preprocedural planning, fabricating in vitro setups, or building a tissue for implantation. Segmentation of tissue types is currently available with standard clinical imaging modalities, and therefore, the lack of use for multimaterial printers is largely due to the expense of current multimaterial printers, and the limitations in the 3D printing technology. There are two classes of improvements that are needed to address these limitations: (1) improved materials that have mechanics tailored for their specific use (e.g., contractile tissue, leaflet, arteries), and (2) improved printers that have the ability to fabricate these

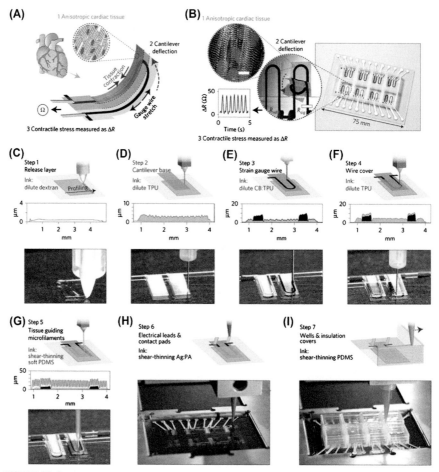

FIGURE 10.8

Multimaterial 3D printer for fabricating drug-testing device for cardiac tissue in vitro. (A) Schematic of cantilever-based tissue contraction assay. (B) Images of cardiac cells and final device. (C–I) Schematics and images of each step of fabrication of the multimaterial device. *Ag*, gold, *CB*, carbon black, *PA*, polyamide; *PDMS*, polydimethylsiloxane, *TPU*, thermoplastic polyurethane.

Reproduced with permission from Lind JU, et al. Instrumented cardiac microphysiological devices via multi-material three-dimensional printing. Nat Mater 2017;16:303–8.

structures that span many orders of magnitude (microns for a single layer of endothelium, to centimeters of thick tissue in the myocardium). With the recent advances in 3D printing hardware, the main limiter seems to be materials development, due to the complexity of developing robust materials that can be utilized in a manner facilitated by 3D printing. More information on these advances can be read in Chapters 3,

4, and 13 of this book. As these improvements are made, however, the commercial use of 3D printing for cardiovascular medicine will go beyond models for preprocedural planning, and into therapies using bioprinted tissue and patient-specific devices.

REFERENCES

[1] Cho GS, Fernandez L, Kwon C. Regenerative medicine for the heart: perspectives on stem-cell therapy. Antioxid Redox Sign 2014;21(14):2018–31. https://doi.org/10.1089/ars.2014.6063. PMID: WOS:000343647700007.

[2] Cohen S, Leor J. Rebuilding broken hearts. Sci Am 2004;291(5):44–51. https://doi.org/10.1038/scientificamerican1104-44. PMID: WOS:000224522300026.

[3] Stevens KR, Kreutziger KL, Dupras SK, Korte FS, Regnier M, Muskheli V, Nourse MB, Bendixen K, Reinecke H, Murry CE. Physiological function and transplantation of scaffold-free and vascularized human cardiac muscle tissue. Proc Natl Acad Sci USA 2009;106(39):16568–73. https://doi.org/10.1073/pnas.0908381106. PMID: WOS:000270305800008.

[4] Giannopoulos AA, Mitsouras D, Yoo SJ, Liu PP, Chatzizisis Y, Rybicki FJ. Applications of 3D printing in cardiovascular diseases. Nat Rev Cardiol 2016;13(12):701–18. https://doi.org/10.1038/nrcardio.2016.170. PMID: WOS:000388584900002.

[5] Tulloch NL, Muskheli V, Razumova MV, Korte FS, Regnier M, Hauch KD, Pabon L, Reinecke H, Murry CE. Growth of engineered human myocardium with mechanical loading and vascular coculture. Circ Res 2011;109(1):47–U195. https://doi.org/10.1161/Circresaha.110.237206. PMID: WOS:000291980200008.

[6] Claes E, Atienza JM, Guinea GV, Rojo FJ, Bernal JM, Revuelta JM, Elices M. Mechanical properties of human coronary arteries. IEEE Eng Med Bio 2010:3792–5. https://doi.org/10.1109/Iembs.2010.5627560. PMID: WOS:000287964004047.

[7] Golob M, Moss RL, Chesler NC. Cardiac tissue structure, properties, and performance: a materials science perspective. Ann Biomed Eng 2014;42(10):2003–13. https://doi.org/10.1007/s10439-014-1071-z. PMID: WOS:000341908500001.

[8] Hasan A, Ragaert K, Swieszkowski W, Selimovic S, Paul A, Camci-Unal G, Mofrad MRK, Khademhosseini A. Biomechanical properties of native and tissue engineered heart valve constructs. J Biomech 2014;47(9):1949–63. https://doi.org/10.1016/j.jbiomech.2013.09.023. PMID: WOS:000338621900004.

[9] Lueders C, Jastram B, Hetzer R, Schwandt H. Rapid manufacturing techniques for the tissue engineering of human heart valves. Eur J Cardiothorac 2014;46(4):593–601. https://doi.org/10.1093/ejcts/ezt510. PMID: WOS:000344968500003.

[10] Vidim K, Wang S-P, Ragan-Kelley J, Matusik W. OpenFab: a programmable pipeline for multi-material fabrication. ACM Trans Graph 2013;32(4):1–12. https://doi.org/10.1145/2461912.2461993.

[11] Sommer K, Izzo RL, Shepard L, Podgorsak AR, Rudin S, Siddiqui AH, Wilson MF, Angel E, Said Z, Springer M, Ionita CN. Design optimization for accurate flow simulations in 3D printed vascular phantoms derived from computed tomography angiography. Proc SPIE Int Soc Opt Eng 2017;10138:101380R. https://doi.org/10.1117/12.2253711. PMID: PMC5485824.

[12] He Y, Yang FF, Zhao HM, Gao Q, Xia B, Fu JZ. Research on the printability of hydrogels in 3D bioprinting. Sci Rep 2016;6. https://doi.org/10.1038/srep29977. Article No. 29977. PMID: WOS:000380056100001.

[13] Lee JM, Sing SL, Tan EYS, Yeong WY. Bioprinting in cardiovascular tissue engineering: a review. Int J Bioprint 2016;2(2):2016. https://doi.org/10.18063/IJB.2016.02.006.

[14] Olivieri LJ, Krieger A, Loke YH, Nath DS, Kim PC, Sable CA. Three-dimensional printing of intracardiac defects from three-dimensional echocardiographic images: feasibility and relative accuracy. J Am Soc Echocardiogr 2015;28(4):392–7. https://doi.org/10.1016/j.echo.2014.12.016. Epub 2015/02/11, 25660668.

[15] Biglino G, Capelli C, Koniordou D, Robertshaw D, Leaver LK, Schievano S, Taylor AM, Wray J. Use of 3D models of congenital heart disease as an education tool for cardiac nurses. Congenit Heart Dis 2017;12(1):113–8. https://doi.org/10.1111/chd.12414. PMID: WOS:000394844000015.

[16] Biglino G, Moharem-Elgamal S, Lee M, Tulloh R, Caputo M. The perception of a three-dimensional-printed heart model from the perspective of different stakeholders: a complex case of truncus arteriosus. Front Pediatr 2017;5:209. https://doi.org/10.3389/fped.2017.00209. PMCID: PMC5626947. Epub 2017/10/17, 29034225.

[17] Wang Z, Liu Y, Luo H, Gao C, Zhang J, Dai Y. Is a three-dimensional printing model better than a traditional cardiac model for medical education? A pilot randomized controlled study. Acta Cardiol Sin 2017;33(6):664–9. https://doi.org/10.6515/acs20170621a. PMCID: PMC5694932. Epub 2017/11/24, 29167621.

[18] Jones TW, Seckeler MD. Use of 3D models of vascular rings and slings to improve resident education. Congenit Heart Dis 2017;12(5):578–82. https://doi.org/10.1111/chd.12486. PMID: WOS:000412170800004.

[19] Garcia J, Yang Z, Mongrain R, Leask RL, Lachapelle K. 3D printing materials and their use in medical education: a review of current technology and trends for the future. BMJ Simul Technol Enhanc Learn 2018;4(1):27–40. https://doi.org/10.1136/bmjstel-2017-000234. PMCID: PMC5765850. Epub 2018/01/23, 29354281.

[20] Jacobs S, Grunert R, Mohr FW, Falk V. 3D-Imaging of cardiac structures using 3D heart models for planning in heart surgery: a preliminary study. Interact Cardiovasc Thorac Surg 2008;7(1):6–9. https://doi.org/10.1510/icvts.2007.156588. Epub 2007/10/11, 17925319.

[21] Valverde I, Gomez G, Suarez-Mejias C, Hosseinpour A-R, Hazekamp M, Roest A, Vazquez-Jimenez JF, El-Rassi I, Uribe S, Gomez-Cia T. 3D printed cardiovascular models for surgical planning in complex congenital heart diseases. J Cardiovasc Magn Reson 2015;17(Suppl. 1):P196. https://doi.org/10.1186/1532-429X-17-S1-P196. PMID: PMC4328535.

[22] Al Jabbari O, Abu Saleh WK, Patel AP, Igo SR, Reardon MJ. Use of three-dimensional models to assist in the resection of malignant cardiac tumors. J Card Surg 2016;31(9):581–3. https://doi.org/10.1111/jocs.12812.

[23] Little SH, Vukicevic M, Avenatti E, Ramchandani M, Barker CM. 3D printed modeling for patient-specific mitral valve intervention: repair with a clip and a plug. JACC Cardiovasc Interv 2016;9(9):973–5. https://doi.org/10.1016/j.jcin.2016.02.027. Epub 2016/05/07, 27151611.

[24] Izzo RL, O'Hara RP, Iyer V, Hansen R, Meess KM, Nagesh SVS, Rudin S, Siddiqui AH, Springer M, Ionita CN. 3D printed cardiac phantom for procedural planning of a transcatheter native mitral valve replacement. Proc SPIE Int Soc Opt Eng 2016;9789,

978908. https://doi.org/10.1117/12.2216952. Medical Imaging 2016: Pacs and Imaging Informatics: Next Generation and Innovations. PMID: WOS:000378538300005.

[25] Mahmood F, Owais K, Taylor C, Montealegre-Gallegos M, Manning W, Matyal R, Khabbaz KR. Three-dimensional printing of mitral valve using echocardiographic data. JACC Cardiovasc Imaging 2015;8(2):227–9. https://doi.org/10.1016/j.jcmg.2014.06.020. Epub 2014/12/03, 25457770.

[26] Vukicevic M, Puperi DS, Jane Grande-Allen K, Little SH. 3D printed modeling of the mitral valve for catheter-based structural interventions. Ann Biomed Eng 2017;45(2):508–19. https://doi.org/10.1007/s10439-016-1676-5. Epub 2016/06/22, 27324801.

[27] Vukicevic M, Avenatti E, Little S. 3D printed patient-specific multi-material modeling of mitral valve apparatus. J Am Coll Cardiol 2017;69(11):1128. PMID: WOS:000397342301650.

[28] McCarthy KP, Ring L, Rana BS. Anatomy of the mitral valve: understanding the mitral valve complex in mitral regurgitation. Eur J Echocardiogr 2010;11(10):i3–9. https://doi.org/10.1093/ejechocard/jeq153. Epub 2010/11/17, 21078837.

[29] Angelini A, Ho SY, Anderson RH, Davies MJ, Becker AE. A histological study of the atrioventricular junction in hearts with normal and prolapsed leaflets of the mitral valve. Br Heart J 1988;59(6):712–6. PMID: 3395530; PMCID: PMC1276881. Epub 1988/06/01.

[30] Akhtar S, Meek KM, James V. Immunolocalization of elastin, collagen type I and type III, fibronectin, and vitronectin in extracellular matrix components of normal and myxomatous mitral heart valve chordae tendineae. Cardiovasc Pathol 1999;8(4):203–11. Epub 2000/03/21, 10724524.

[31] Zuo KP, Pham T, Li KW, Martin C, He ZM, Sun W. Characterization of biomechanical properties of aged human and ovine mitral valve chordae tendineae. J Mech Behav Biomed 2016;62:607–18. https://doi.org/10.1016/j.jmbbm.2016.05.034. PMID: WOS:000381238500053.

[32] Stella JA, Sacks MS. On the biaxial mechanical properties of the layers of the aortic valve leaflet. J Biomech Eng 2007;129(5):757–66. https://doi.org/10.1115/1.2768111. Epub 2007/09/25, 17887902.

[33] Butcher JT, Nerem RM. Porcine aortic valve interstitial cells in three-dimensional culture: comparison of phenotype with aortic smooth muscle cells. J Heart Valve Dis 2004;13(3):478–85. discussion 85-6. Epub 2004/06/30, 15222296.

[34] Butcher JT, Simmons CA, Warnock JN. Mechanobiology of the aortic heart valve. J Heart Valve Dis 2008;17(1):62–73. Epub 2008/03/28, 18365571.

[35] Duan B, Hockaday LA, Kang KH, Butcher JT. 3D bioprinting of heterogeneous aortic valve conduits with alginate/gelatin hydrogels. J Biomed Mater Res 2013;101(5):1255–64. https://doi.org/10.1002/jbm.a.34420. PMCID: PMC3694360. Epub 2012/09/28, 23015540.

[36] Duan B, Hockaday LA, Kapetanovic E, Kang KH, Butcher JT. Stiffness and adhesivity control aortic valve interstitial cell behavior within hyaluronic acid based hydrogels. Acta Biomater 2013;9(8):7640–50. https://doi.org/10.1016/j.actbio.2013.04.050. PMCID: PMC3700637. Epub 2013/05/08, 23648571.

[37] Duan B, Kapetanovic E, Hockaday LA, Butcher JT. Three-dimensional printed trileaflet valve conduits using biological hydrogels and human valve interstitial cells. Acta Biomater 2014;10(5):1836–46. https://doi.org/10.1016/j.actbio.2013.12.005. PMCID: PMC3976766. Epub 2013/12/18, 24334142.

[38] Hockaday LA, Duan B, Kang KH, Butcher JT. 3D-printed hydrogel technologies for tissue-engineered heart valves. 3D Print Addit Manuf 2014;1(3):122–36. https://doi.org/10.1089/3dp.2014.0018. PMID: WOS:000209657800004.

[39] Hockaday LA, Kang KH, Colangelo NW, Cheung PY, Duan B, Malone E, Wu J, Girardi LN, Bonassar LJ, Lipson H, Chu CC, Butcher JT. Rapid 3D printing of anatomically accurate and mechanically heterogeneous aortic valve hydrogel scaffolds. Biofabrication 2012;4(3):035005. https://doi.org/10.1088/1758-5082/4/3/035005. PMCID: PMC3676672. Epub 2012/08/24, 22914604.

[40] Kang LH, Armstrong PA, Lee LJ, Duan B, Kang KH, Butcher JT. Optimizing photo-encapsulation viability of heart valve cell types in 3D printable composite hydrogels. Ann Biomed Eng 2017;45(2):360–77. https://doi.org/10.1007/s10439-016-1619-1. PMCID: PMC5075276. Epub 2016/04/24, 27106636.

[41] Nakayama Y, Takewa Y, Sumikura H, Yamanami M, Matsui Y, Oie T, Kishimoto Y, Arakawa M, Ohmuma K, Tajikawa T, Kanda K, Tatsumi E. In-body tissue-engineered aortic valve (biovalve type VII) architecture based on 3D printer molding. J Biomed Mater Res B Appl Biomater 2015;103(1):1–11. https://doi.org/10.1002/jbm.b.33186. Epub 2014/04/26, 24764308.

[42] Hillis LD, Smith PK, Anderson JL, Bittl JA, Bridges CR, Byrne JG, Cigarroa JE, DiSesa VJ, Hiratzka LF, Hutter Jr AM, Jessen ME, Keeley EC, Lahey SJ, Lange RA, London MJ, Mack MJ, Patel MR, Puskas JD, Sabik JF, Selnes O, Shahian DM, Trost JC, Winniford MD, Jacobs AK, Anderson JL, Albert N, Creager MA, Ettinger SM, Guyton RA, Halperin JL, Hochman JS, Kushner FG, Ohman EM, Stevenson W, Yancy CW. 2011 ACCF/AHA guideline for coronary artery bypass graft surgery: executive summary: a report of the American College of Cardiology Foundation/American Heart Association Task Force on Practice Guidelines. J Thorac Cardiovasc Surg 2012;143(1):4–34. https://doi.org/10.1016/j.jtcvs.2011.10.015. Epub 2011/12/17, 22172748.

[43] Desai M, Seifalian AM, Hamilton G. Role of prosthetic conduits in coronary artery bypass grafting. Eur J Cardiothorac 2011;40(2):394–8. https://doi.org/10.1016/j.ejcts.2010.11.050. PMID: WOS:000292690200029.

[44] Seifu DG, Purnama A, Mequanint K, Mantovani D. Small-diameter vascular tissue engineering. Nat Rev Cardiol 2013;10(7):410–21. https://doi.org/10.1038/nrcardio.2013.77. Epub 2013/05/22, 23689702.

[45] Chen QZ, Bismarck A, Hansen U, Junaid S, Tran MQ, Harding SE, Ali NN, Boccaccini AR. Characterisation of a soft elastomer poly(glycerol sebacate) designed to match the mechanical properties of myocardial tissue. Biomaterials 2008;29(1):47–57. https://doi.org/10.1016/j.biomaterials.2007.09.010. WOS:000251210600005.

[46] Waller BF, Orr CM, Slack JD, Pinkerton CA, Van Tassel J, Peters T. Anatomy, histology, and pathology of coronary arteries: a review relevant to new interventional and imaging techniques—Part I. Clin Cardiol 1992;15(6):451–7. Epub 1992/06/01. PMID: 1617826.

[47] Ong CS, Zhou X, Huang CY, Fukunishi T, Zhang HT, Hibino N. Tissue engineered vascular grafts: current state of the field. Expert Rev Med Device 2017;14(5):383–92. https://doi.org/10.1080/17434440.2017.1324293. PMID: WOS:000401466800007.

[48] Yang Y, Liu X, Xia Y, Liu X, Wu W, Xiong H, Zhang H, Xu L, Wong KKL, Ouyang H, Huang W. Impact of spatial characteristics in the left stenotic coronary artery on the hemodynamics and visualization of 3D replica models. Sci Rep 2017;7(1):15452.

https://doi.org/10.1038/s41598-017-15620-1. PMCID: PMC5684364. Epub 2017/11/15, 29133915.

[49] Kolesky DB, Truby RL, Gladman AS, Busbee TA, Homan KA, Lewis JA. 3D bioprinting of vascularized, heterogeneous cell-laden tissue constructs. Adv Mater 2014;26(19):3124−30. https://doi.org/10.1002/adma.201305506. Epub 2014/02/20, 24550124.

[50] Cui X, Boland T. Human microvasculature fabrication using thermal inkjet printing technology. Biomaterials 2009;30(31):6221−7. https://doi.org/10.1016/j.biomaterials.2009.07.056. Epub 2009/08/22, 19695697.

[51] Mosadegh B, Xiong G, Dunham S, Min JK. Current progress in 3D printing for cardiovascular tissue engineering. Biomed Mat 2015;10(3):034002. https://doi.org/10.1088/1748-6041/10/3/034002. Epub 2015/03/17, 25775166.

[52] Cambria E, Pasqualini FS, Wolint P, Gunter J, Steiger J, Bopp A, Hoerstrup SP, Emmert MY. Translational cardiac stem cell therapy: advancing from first-generation to next-generation cell types. NPJ Regen Med 2017;2:17. https://doi.org/10.1038/s41536-017-0024-1. PMCID: PMC5677990. Epub 2018/01/06, 29302353.

[53] Zhang J. Engineered tissue patch for cardiac cell therapy. Curr Treat Options Cardiovasc Med 2015;17(8):399. https://doi.org/10.1007/s11936-015-0399-5. PMID: PMC4676725.

[54] Rienks M, Papageorgiou AP, Frangogiannis NG, Heymans S. Myocardial extracellular matrix: an ever-changing and diverse entity. Circ Res 2014;114(5):872−88. https://doi.org/10.1161/CIRCRESAHA.114.302533. Epub 2014/03/01, 24577967.

[55] Gaetani R, Doevendans PA, Metz CH, Alblas J, Messina E, Giacomello A, Sluijter JP. Cardiac tissue engineering using tissue printing technology and human cardiac progenitor cells. Biomaterials 2012;33(6):1782−90. https://doi.org/10.1016/j.biomaterials.2011.11.003. Epub 2011/12/06, 22136718.

[56] Hernandez-Cordova R, Mathew DA, Balint R, Carrillo-Escalante HJ, Cervantes-Uc JM, Hidalgo-Bastida LA, Hernandez-Sanchez F. Indirect three-dimensional printing: a method for fabricating polyurethane-urea based cardiac scaffolds. J Biomed Mater Res 2016;104(8):1912−21. https://doi.org/10.1002/jbm.a.35721. PMCID: PMC5338726. Epub 2016/03/19, 26991636.

[57] Ho CM, Mishra A, Lin PT, Ng SH, Yeong WY, Kim YJ, Yoon YJ. 3D printed polycaprolactone carbon nanotube composite scaffolds for cardiac tissue engineering. Macromol Biosci 2017;17(4). https://doi.org/10.1002/mabi.201600250. Epub 2016/11/29, 27892655.

[58] Izadifar M, Chapman D, Babyn P, Chen X, Kelly ME. UV-assisted 3D bioprinting of nanoreinforced hybrid cardiac patch for myocardial tissue engineering. Tissue Eng Part C Methods 2018;24(2):74−88. https://doi.org/10.1089/ten.TEC.2017.0346. Epub 2017/10/21, 29050528.

[59] Wang Z, Lee SJ, Cheng HJ, Yoo JJ, Atala A. 3D bioprinted functional and contractile cardiac tissue constructs. Acta Biomater 2018. https://doi.org/10.1016/j.actbio.2018.02.007. Epub 2018/02/17, 29452273.

[60] Lind JU, Busbee TA, Valentine AD, Pasqualini FS, Yuan H, Yadid M, Park SJ, Kotikian A, Nesmith AP, Campbell PH, Vlassak JJ, Lewis JA, Parker KK. Instrumented cardiac microphysiological devices via multimaterial three-dimensional printing. Nat Mater 2017;16(3):303−8. https://doi.org/10.1038/nmat4782. PMCID: PMC5321777. Epub 2016/11/01, 27775708.

[61] Giannopoulos AA, Steigner ML, George E, Barile M, Hunsaker AR, Rybicki FJ, Mitsouras D. Cardiothoracic applications of 3-dimensional printing. J Thorac Imag 2016;31(5):253−72. https://doi.org/10.1097/rti.0000000000000217. PMCID: PMC4993676. Epub 2016/05/06, 27149367.

[62] Yang DH, Kang JW, Kim N, Song JK, Lee JW, Lim TH. Myocardial 3-dimensional printing for septal myectomy guidance in a patient with obstructive hypertrophic cardiomyopathy. Circulation 2015;132(4):300−1. https://doi.org/10.1161/circulatio-naha.115.015842. Epub 2015/07/29, 26216088.

[63] Vukicevic M, Mosadegh B, Min JK, Little SH. Cardiac 3D printing and its future directions. JACC Cardiovasc Imaging 2017;10(2):171−84. https://doi.org/10.1016/j.jcmg.2016.12.001. PMID: WOS:000394926500011.

[64] Maragiannis D, Jackson MS, Igo SR, Chang SM, Zoghbi WA, Little SH. Functional 3D printed patient-specific modeling of severe aortic stenosis. J Am Coll Cardiol 2014;64(10):1066−8. https://doi.org/10.1016/j.jacc.2014.05.058. Epub 2014/09/06, 25190245.

[65] Maragiannis D, Jackson MS, Igo SR, Schutt RC, Connell P, Grande-Allen J, Barker CM, Chang SM, Reardon MJ, Zoghbi WA, Little SH. Replicating patient-specific severe aortic valve stenosis with functional 3D modeling. Circ Cardiovasc Imaging 2015;8(10):e003626. https://doi.org/10.1161/circimaging.115.003626. Epub 2015/10/10, 26450122.

[66] Jang J, Park HJ, Kim SW, Kim H, Park JY, Na SJ, Kim HJ, Park MN, Choi SH, Park SH, Kim SW, Kwon SM, Kim PJ, Cho DW. 3D printed complex tissue construct using stem cell-laden decellularized extracellular matrix bioinks for cardiac repair. Biomaterials 2017;112:264−74. https://doi.org/10.1016/j.biomaterials.2016.10.026. PMID: WOS: 000389166700023.

Assessing Perfusion Using 3D Bioprinting

11

Hanley Ong[1,2], Kranthi K. Kolli[1,2]

Department of Radiology, Weill Cornell Medicine, New York, NY, United States[1]; Dalio Institute of Cardiovascular Imaging, NewYork-Presbyterian Hospital, New York, NY, United States[2]

3D BIOPRINTING OF PERFUSABLE VASCULAR NETWORKS

Living tissues have an indigenous mass transport system that involves vascular networks, which are capable of facilitating nutrient delivery, gas exchange, and waste disposal. Tissues outside of a certain proximity to a vascular network ($\sim 150-200$ μm, the diffusion depth of oxygen) rapidly lose cellular structure and function, eventually developing central necrotic cores [1,2]. This progressive degeneration is particularly pronounced in highly metabolic tissues, such as cardiovascular tissue. One of the central challenges of tissue engineering has thus been ensuring that the engineered tissue is adequately perfused by a functioning vascular network with flow, as absence of such a network limits the thickness of tissue that can be fabricated to a few hundred micrometers [3,4].

Broadly speaking, there are two processes by which vascular networks can be generated: vasculogenesis (de novo formation of blood vessels) and angiogenesis (sprouting of new vessels from existing vessels). Most traditional tissue engineering methods, such as micromolding, biotextile application, and photolithography, have tackled the issue of perfusing engineered tissue by utilizing proangiogenic factors to induce host-generated angiogenesis [5–8]. However, the time taken for new blood vessels to form exceeds the time in which tissue necrosis occurs, thus limiting in vivo cell viability. Additionally, the cost of utilizing angiogenic factors to vascularize large-scale tissue constructs is often prohibitively high [1,2].

In order to tackle the above limitations, researchers have begun to supplement traditional tissue engineering methods with three-dimensional (3D) bioprinting, which refers to the deposition of living cells, biomaterials, and scaffolds in a layer-by-layer fashion into 3D constructs [9]. Although this process is inherently more complex than other cell-free tissue engineering methods, such as traditional 3D printing (e.g., due to the need for the printed materials to be biocompatible), 3D bioprinting offers several advantages as will be described in detail below. 3D bioprinting presents a unique way to prevascularize tissue constructs in vitro, allowing for improved cell viability in vivo [2]. Furthermore, 3D bioprinting also

3D Printing Applications in Cardiovascular Medicine. https://doi.org/10.1016/B978-0-12-803917-5.00011-0

allows unprecedented control over the types of materials and cells used, and enables these materials to be deposited in precise locations so as to replicate the native tissue architecture [1]. It should be noted that the current 3D bioprinting approaches have not entirely solved the challenge of fabricating thicker, well-perfused tissues plaguing traditional tissue engineering. Recent advances in 3D bioprinting technologies have, however, made it increasingly possible to engineer tissues of larger scale and greater complexity in a high-throughput and controlled fashion.

MATERIALS AND METHODS

3D bioprinting is the process of additively depositing bioink, which consists of living cells, biomaterials, and structural scaffolds, onto a surface to produce 3D structures. When used in conjunction with medical imaging modalities, it is possible to construct biomimetic physical models of patient tissues [10]. There exist many different types of bioprinters, but the most common are laser-based, inkjet-based, and extrusion-based (Fig. 11.1). Laser-based bioprinting harnesses the energy from focused laser pulses to form high-pressure bubbles on a cell-laden donor surface, which when ruptured, generates shock waves that propel individual cells towards the printing surface. Inkjet-based printers deposit droplets of bioink through nozzles, and fall into two categories: thermal and piezoelectric. The former uses heat to generate a bubble that propels ink through the nozzle, while the latter applies electric current to a piezoelectric crystal that vibrates in response and forces bioink through the nozzle. Extrusion-based printers also use a nozzle to deposit bioink, but rely on pressure (via a pneumatic or mechanical pump) to push the

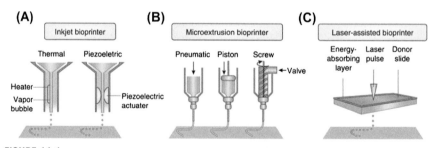

FIGURE 11.1

Common 3D bioprinters. Visual depiction of common 3D bioprinters. (A) Inkjet bioprinters typically fall into one of two categories: (1) thermal bioprinters, which use heat to generate a bubble that forces ink through the nozzle, and (2) piezoelectric bioprinters, which apply electrical currents to a piezoelectric crystal and cause it to vibrate, propelling ink through the nozzle. (B) Extrusion bioprinters use pressure generated through a variety of means (e.g., pneumatic pump, piston, screw) to deposit ink. (C) Laser bioprinters harness energy from laser pulses to propel individual cells from a donor slide onto the printing surface.

Reproduced with permission from Murphy SV, et al. 3D bioprinting of tissues and organs. Nat Biotechnol 2014;
32:773–85.

Table 11.1 Advantages/Disadvantages of Common Bioprinters [11–13,30]

Printer	Advantages	Disadvantages
Laser	• Deposit cells at high density (10^8 cells/mL) • High resolution (single cells) • Can print high viscosity bioinks	• Time-consuming to prepare cell-laden donor surface • Slow printing speed • Heat from laser energy may damage cells • High cost
Inkjet	• Create concentration gradient of cells, materials, growth factors by altering drop density/size during printing • High speed • Cost effective ("drop-on-demand," utilize exact amount of material needed)	• Exposes cells to thermal/mechanical stress • Nozzle clogging (especially for high-viscosity bioink) • Requires liquid bioink that must solidify after printing • Lower cell density
Extrusion	• Can modify extrusion pressure to print high viscosity bioink • Cost effective • High cell density	• Nozzle clogging (especially for high-viscosity bioink) • Exposes cells to shear stress at high pressures → lower cell viability

bioink through [10–13]. Each of these bioprinters has distinct advantages and disadvantages, as summarized in Table 11.1.

More recently, using these different types of bioprinters (largely nozzle-based), researchers have developed various methods for fabricating vascular networks, with varying degrees of success [2]. Two general strategies have emerged for the fabrication of vascular networks: additive methods, in which biomaterials are directly printed layer by layer into the eventual 3D construct; and subtractive methods, in which a soluble material is first constructed into a sacrificial template, then cast in hydrogel containing cells and biomaterials, and finally solubilized to reveal the resulting vascular network. Early attempts at additively bioprinting vasculature mostly relied on extrusion bioprinting, which involved long cell handling times and application of high shear stresses that could damage the cells. Consequently, these methods were often slow and limited in print resolution [1,3]. Many of the early subtractive bioprinting attempts also had their own flaws. Some studies implemented techniques commonly used in the semiconductor industry (e.g., photolithography, thin film deposition, etc.) and successfully demonstrated cell viability up to 200 µm away from the fabricated vasculature [14,15]. However, construction of these channels required delicately aligning and stacking thin layers of planar networks, which was highly impractical for tissue engineering applications [3,16,17]. Other methods of implementing sacrificial templates required use of cytotoxic materials in dissolving the template or casting the channels, thus precluding in vivo application [3].

INNOVATIONS IN SUBTRACTIVE 3D BIOPRINTING OF VASCULARIZED TISSUES

Recently, Miller et al. developed a seminal subtractive method for vasculature fabrication in which they utilized biocompatible carbohydrate glass as a printing template [3]. This novel carbohydrate glass was formed by dissolving a mixture of carbohydrates (glucose, sucrose, and dextran) in water, then boiling off the solvent; the carbohydrate glass possessed sufficient mechanical rigidity to support its own weight during the casting process, which was an issue with biocompatible sacrificial materials used previously. Miller et al. first formed the carbohydrate glass into a self-supporting lattice, and then coated it with a homogeneous suspension of cells in extracellular matrix (ECM). They demonstrated ECM cross-linkage with a variety of matrix materials (including fibrin, agarose, Matrigel), and after allowing the ECM to cross-link, they dissolved the glass template with aqueous solution and allowed it to flow through the newly formed channels (Fig. 11.2A). Finally, they seeded these channels with endothelial cells (ECs), creating nonleaking, perfusable vasculature. They then demonstrated the ability of these channels to perfuse tissue and prevent necrosis, measuring green fluorescent protein expression in cell-laden slab gels with and without channels inserted. When submersed in media, the gels without channels exhibited cellular activity only along the periphery where diffusion of nutrients could sustain cell growth, while the gels with channels exhibited cellular activity at the core due to adequate perfusion from the channels (Fig. 11.2B), thus illustrating this method's ability to support survival of thicker engineered tissues previously limited by diffusional constraints. Importantly, it was reported that casting bioink around the glass lattices allowed rapid formation of perfusable channels, enabling much faster perfusion of the tissue than was previously possible, and thus saving time and preventing formation of a necrotic core during tissue fabrication.

Although Miller et al.'s study represents an important step forward in the field of tissue engineering, the vascular network it fabricated was limited in thickness (<1 mm) and culture times (<14 days) due to a reliance on convective media flow to perfuse the tissue [18]. This inability to directly perfuse the vasculature and surrounding tissue was addressed in several later studies as discussed below. Lee et al. described an innovative subtractive 3D bioprinting method that used only cells and biomaterials during the printing process, in contrast to previous methodologies that utilized synthetic materials (e.g., plastics, polydimethylsiloxane [PDMS]) [19]. Using collagen precursor, gelatin, and human umbilical vein endothelial cells (HUVECs), they printed a vascular channel through a layer-by-layer approach (Fig. 11.3). First, layers of collagen were printed in a flow chamber to form a base. Next, a 1:1 mixture of gelatin and HUVECs was printed in a straight line on top of the collagen base. Collagen was then printed on both sides and on top of this HUVEC/gelatin rod. The flow chamber was sealed and incubation of the printed structure at 37°C led to collagen cross-linking and gelatin liquefaction, resulting in the HUVECs within the rod attaching along the inner surface of the

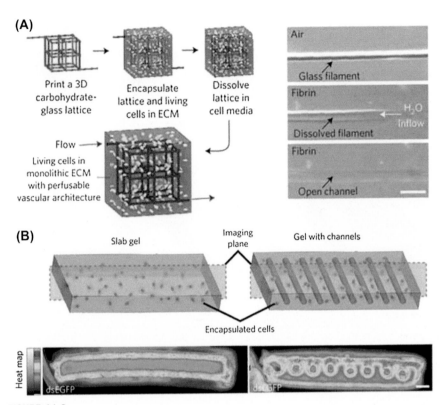

FIGURE 11.2

(A) 3D bioprinting described by Miller et al. Visual depiction of Miller et al.'s bioprinting process. A carbohydrate glass lattice is created and coated with bioink. Then, the carbohydrate lattice is dissolved and the extracellular matrix (ECM) within the bioink cross-links and solidifies, forming vascular channels. (B) Cross-sections of cell-laden slab gels with and without vascular channels. The gel with vascular channels exhibits markedly higher dsEGFP (destabilized enhanced green fluorescent protein) expression (and thus, cell activity) within the gel, while the gel without vascular channels only exhibits expression along the periphery.

Reproduced with permission from Miller JS, et al. Rapid casting of patterned vascular networks for perfusable engineered three-dimensional tissues. Nat Mater 2012;11:768–74.

newly formed collagen channel. Finally, the liquefied gelatin was washed out with gentle media flow directly applied to the flow chamber, resulting in a perfused vascular channel. The ability of the channel to support adjacent tissue was assessed by seeding the collagen with varying densities of ECs, while the ability of the channel to prevent leakage was assessed by injecting fluorescently tagged bovine serum albumin and dextran in the media. It was also reported that the fabricated vascular channel could guarantee high cell viability ($\sim$90%) and barrier function in tissues of

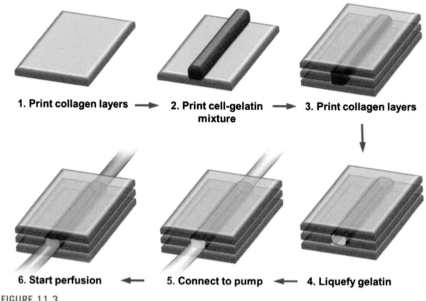

1. Print collagen layers → 2. Print cell-gelatin mixture → 3. Print collagen layers

↓

6. Start perfusion ← 5. Connect to pump ← 4. Liquefy gelatin

FIGURE 11.3

3D bioprinting method described by Lee et al. Visual depiction of Lee et al.'s bioprinting process. Layers of collagen are deposited to form a base layer, upon which a rod of 1:1 gelatin and human umbilical vein endothelial cells mixture is printed. Additional collagen is printed around the cell-gelatin rod, the flow chamber is sealed, and the gelatin is liquefied. Finally, the chamber is connected to a pump and perfused.

Reproduced with permission from Lee VK, et al. Creating perfused functional vascular channels using 3D bio-printing technology. Biomaterials 2014;35:8092–102.

up to 5 mm in thickness and 5 million cells/mL density for several weeks. Lee et al.'s work provided a way to generate thicker, complex vascular networks with longer lifetimes from biocompatible materials, marking a shift from previous tissue engineering efforts that utilized nonbiocompatible materials and involved cumbersome, multistep processing [19].

The recent work of Kolesky et al. also represented another step forward in extending culture lifetimes (>6 weeks) and thickness (>1 cm) of engineered tissues through 3D bioprinting of directly perfusable vascular networks [18]. Moreover, the authors took advantage of the longer culture lifetimes of their vascularized tissues and perfused them with growth factors, demonstrating an ability to promote cellular differentiation in engineered tissue. Their bioprinting method involved the use of several bioinks: a cell-laden ink, a fugitive (sacrificial) ink, and a castable ECM (Fig. 11.4). The cell-laden and fugitive inks were first printed into a network of channels in a silicone perfusion chip. After casting ECM material over this network, the fugitive ink was then removed by inducing a gel-to-fluid transition in the triblock copolymer making up the fugitive ink [polyethylene oxide (PEO)-propylene oxide (PPO)-PEO]. The ECM material and cell-laden inks were composed of a

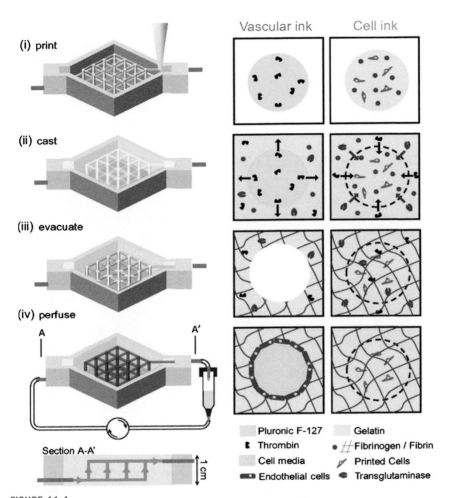

FIGURE 11.4

3D bioprinting method described by Kolesky et al. Visual depiction of Kolesky et al.'s bioprinting process. Cell-laden and fugitive inks are first printed into a lattice inside a silicone perfusion chip (A). Then, extracellular matrix is cast over the lattice (B). The fugitive ink is evacuated (C), and the resulting vascular network is connected to a perfusion pump (D).

Reproduced with permission from Kolesky DB, et al. Three-dimensional bioprinting of thick vascularized tissues.
Proc Natl Acad Sci 2016;113:3179–84.

fibrin-gelatin mixture that were cross-linked by thrombin and transglutaminase enzymes present in the castable ECM. Finally, the resultant vascular channels were lined with HUVECs and directly perfused via external pump. Kolesky et al. have demonstrated the ability to print a heterogeneous, multimaterial tissue architecture that integrated parenchyma, stroma, and endothelium and thus more accurately represented native tissue: they incorporated human mesenchymal stem cells (hMSCs) in the cell-laden ink, seeded the vasculature with HUVECs, and filled the remaining interstitial space with human neonatal dermal fibroblasts (hNDFs). This tissue was then perfused with osteogenic growth factors (BMP-2, ascorbic acid, glycerophosphate) to successfully induce osteogenic differentiation of the hMSCs over 6 weeks [18]. The breakthroughs in this study constitute significant developments in the effort to engineer complex vascularized tissues.

INNOVATIONS IN ADDITIVE 3D BIOPRINTING OF VASCULARIZED TISSUES

One drawback of the subtractive methods as discussed previously has been the inherent complexity involved in multistep bioprinting, which may limit the size, morphology, and functionality of the fabricated vasculature. Bioprinting of vasculature through single-step processes, such as additive extrusion printing, is thus desirable. Recently, Jia et al. developed a single-step bioprinting method, in which a trilayered coaxial nozzle printer with concentric channels was used to simultaneously deposit a blended bioink consisting of gelatin methacryloyl (GelMA), sodium alginate, and 4-arm poly(ethylene glycol)-tetra-acrylate (PEGTA) [5]. This particular mix of hydrogel materials was chosen for its dual cross-linking ability, which imparted significant mechanical stability to the vascular channels: the alginate enabled rapid ionic cross-linking by calcium ions during initial deposition, while the GelMA and PEGTA ensured longer-term stability through covalent UV light photocross-linking. Overall, this printing method proved to be highly flexible in generation of complex vascular structures: by adjusting the needle size in the nozzle, vascular tubes of varying sizes could be printed; and by depositing the bioink in a single step, vascular tubes of varying shapes could be fabricated much more easily (Fig. 11.5). Cell viability of the surrounding tissue (HUVECs and hMSCs) supported by this vasculature exceeded 80% for up to several weeks in tissue constructs less than 1 cm thick—lower than that observed in sacrificial templating methods, but greater than that observed in previous extrusion-based methods [5].

APPLICATIONS IN TRANSPLANTATION AND REGENERATIVE MEDICINE

Current autologous and homologous transplant treatments for end-stage organ failure are often complicated by problems such as donor-related infection, primary graft dysfunction, rejection, and long-term immunosuppression related infection, thus creating a need for better therapeutic options. One major potential application of bioprinted vascularized tissue is its use as a tissue replacement, enabling novel treatment for tissue damage and end-stage organ failure. Bioprinting holds particular clinical promise in regenerative medicine because it enables fabrication of tissue using patient-specific cells (through the use of pluripotent stem cells), thus

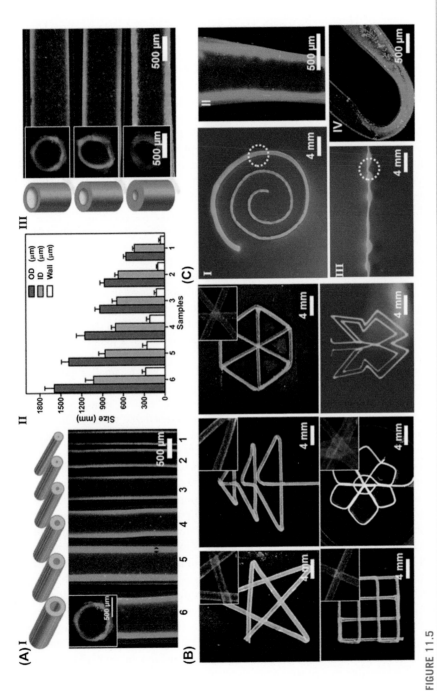

FIGURE 11.5

Demonstration of flexible 3D bioprinting method developed by Jia et al. Several demonstrations of how Jia et al.'s 3D bioprinting method can be used to print vasculature of (A) different thicknesses and (B and C) varying morphologies.

Reproduced with permission from Jia W, et al. Direct 3D bioprinting of perfusable vascular constructs using a blend bioink. Biomaterials 2016;106:58–68.

circumventing many of the issues associated with conventional transplant treatments [20]. The range of bioprintable tissue types being investigated is extensive, and includes cardiac, valvular, osteogenic, liver, lung, cartilaginous, and pancreatic tissues [20].

Currently, it is not clinically feasible to directly transplant bioprinted tissue for therapeutic benefit due to several reasons. Historically, the main problem has been the inability to construct tissues thicker than a few hundred microns due to the absence of adequately perfused vasculature. Another limitation has been the inability to incorporate multiple cell types and materials in the engineered tissue, and to induce differentiation and expression of the proper tissue phenotype. Furthermore, the cell densities that can be generated in fabricated tissues are much lower than cell densities of native tissue, limiting the clinical value of using these fabricated tissues in treating organic pathology (e.g., ischemic damage to myocardial tissue) [21].

With recent advances in bioprinting methodologies and technologies, much of these limitations are gradually being addressed. Researchers have demonstrated the ability to fabricate tissues thicker than 1 cm, incorporate multiple cell types, and induce tissue differentiation with growth factors [18]. Recently, Ball et al. designed a 3D bioprinted vascular network capable of supporting a tissue engineered bone construct of >20 cm^3 in vitro, with the goal of eventually engineering bone grafts for in vivo use [22]. However, for extensively vascularized and metabolically active tissues like those found in the heart, liver, and pancreas, a major barrier to clinical application of bioprinted tissues is the requirement of a hierarchical vascular architecture spanning from arteries and veins down to small capillaries [20]. Bioprinting technologies thus far are unable to fabricate vasculature at the submicron scale, but a possible alternative is to allow native angiogenesis to create the microvasculature after implantation of the vascularized tissue graft.

An additional important consideration is the need for a simple and reliable method of integrating the fabricated tissue into the host. To address this consideration, Sooppan et al. developed a preliminary proof-of-concept technique for in vivo implantation of vascularized tissue, which was tested in rat models [23]. After using sacrificial carbohydrate glass-extrusion bioprinting to fabricate vascularized PDMS gels, they implanted the PDMS gels in line with the femoral arteries of rats. Compared to negative controls in which the femoral arteries were double ligated and transected (and consequently no flow was expected), the femoral arteries ligated to the PDMS gels exhibited significantly greater perfusion up to 3 h after implantation (Fig. 11.6). Unfortunately, flow through the gels was negligible after 3 h due to clotting, signifying a need for improved hemocompatibility and antithrombogenicity in future attempts to implant vascularized tissues.

A later study by Mirabella et al. addressed these issues related to clotting by seeding the bioprinted vasculature with ECs, thus enabling their resulting bioprinted tissue to survive for up to several weeks post implantation in vivo [24]. They

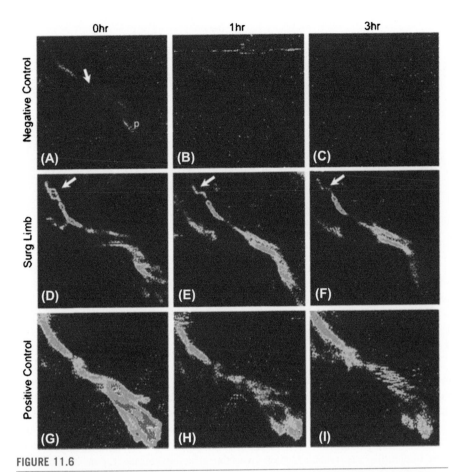

FIGURE 11.6

Doppler imaging studies of rat femoral arteries. (A–C) Negative control: femoral arteries ligated and transected; no blood flow expected. (D–F) Surgical limb: femoral arteries ligated to vascular patch; blood flow through the fabricated vascular tissue is observed. (G–I) Positive control: no ligation of femoral arteries; normal blood flow expected.

Reproduced with permission from Sooppan R, Paulsen SJ, Han J, et al. In vivo anastomosis and perfusion of a three-dimensionally-printed construct containing microchannel networks. Tissue Eng Part C Methods 2016;22: 1–7.

investigated the clinical value of transplanted bioprinted tissues in restoring perfusion to ischemic areas. After encapsulating carbohydrate glass-extrusion bioprinted vascularized tissues in a fibrin patch and seeding the vasculature with ECs, they implanted the vascular patches in a mouse model of hind limb ischemia. In this model, the left femoral artery was occluded, inducing a 50% postoperative decrease in distal perfusion. Mice with the bioprinted vascular patches implanted

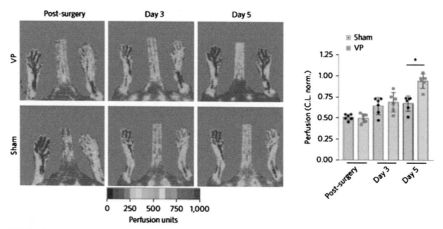

FIGURE 11.7

Doppler imaging studies of mouse model of hind limb ischemia. Following occlusion of the left femoral artery in mice to induce ischemia, some mice were transplanted with vascular patches (VP) while others were not (sham). Doppler imaging shows restoration of perfusion 5 days postoperatively in the VP group, but not in the sham group.

Reproduced with permission from Mirabella T, et al. 3D-printed vascular networks direct therapeutic angiogenesis in ischaemia. Nat Biomed Eng 2017;1:83.

at the injury site exhibited a marked improvement in perfusion, reaching near-normal values within 5 days post implantation, whereas the mice without vascular patches showed no improvement (Fig. 11.7). They also assessed the therapeutic value of their patch in the more clinically urgent situation of a myocardial infarction (MI), inducing MIs in rats by ligating the left anterior descending coronary artery, then implanting the vascular patch over the affected cardiac region. Significant rescue of cardiac function was observed, as shown by the preserved ejection fraction and cardiac output, as well as the increased cardiac capillary density of the rats that received vascular patches [24]. Another important finding reported in this study was that the geometry of the implanted vasculature impacted therapeutic function, such as decreasing vessel diameter from 400 to 200 μm and changing microchannel arrangement from parallel to grid-like; these factors decreased the ability of the vascular patch to rescue perfusion, likely due to diminished efficiency of directing blood flow to ischemic regions. Overall, Mirabella et al. illustrated the clinical value that bioprinted tissues could have in treating ischemic tissue damage, particularly for patients who cannot tolerate traditional open surgical bypass or whose disease affects microvasculature rather than large arteries [24].

Although highly promising, applications of bioprinted vascularized tissue in transplantation medicine do not yet have significant clinical relevance due to the practical considerations discussed above. In recent years, researchers have taken and continue to take incremental steps towards realizing this goal, overcoming the various limitations of tissue engineering along the way.

APPLICATIONS AS MEDICAL RESEARCH MODELS

A second major application of bioprinted vascularized tissues is their use as experimental models to assess perfusion under different pathological and nonpathological states. Because many 3D bioprinting techniques have been developed only fairly recently, 3D bioprinted vascularized tissues are not yet in widespread use as models in medical research. However, it should be noted that this technology has vast future potential, and the relevant studies conducted so far also have yielded promising results.

3D bioprinted tissues can serve as an excellent in vitro model for cancer research, capturing the complex tumor microenvironment more accurately than traditional 2D cell culture systems [20,25]. In particular, the aberrant angiogenesis and blood flow associated with tumorigenesis could be studied in a highly controlled fashion with 3D bioprinted tissue models, allowing for development of more effective cancer therapies [1]. Researchers have demonstrated the ability to fabricate human ovarian cancer cells, bioprint immortal HeLa cells for use in a cervical tumor model, and even assess the effects of chemotherapeutic drug tamoxifen using a breast cancer model [20]. More recently, Trachtenberg et al. used extrusion bioprinting to fabricate multilayered poly(propylene fumarate) scaffolds that were seeded with Ewing Sarcoma cells and connected to perfusion bioreactors [25]. By varying the pore sizes to generate pore gradients in the printed constructs and by modulating flow through the perfusion bioreactors, this method enabled them to re-create the highly heterogeneous permeability gradients and cell environment found in tumors. They demonstrated that tumor cell proliferation and protein expression depended on the pore gradient and shear stress gradients from flow, thus illustrating how the accuracy of their in vitro 3D printed model enabled improved understanding of cancer biology.

3D bioprinted vascularized tissue models would also be of a greater value for research in the field of blood flow and atherosclerosis, in which turbulent or diminished flow leads to EC damage, and consequently, formation of atherosclerotic lesions. Flow rates seem to play an important role in the pathogenesis of atherosclerosis, as recent studies suggest that physiologic flow rates preserve EC function [20]. Use of 3D bioprinted models, in which flow rates and vascular morphologies can be easily modulated, would enable researchers to study atherosclerosis in a more controlled fashion.

Although 3D bioprinted tissues are not yet being widely used to model perfusion in medical research, regular 3D printing has already been used to great effect for this purpose. At the most fundamental level, 3D printed models can be used to better visualize and understand anatomy, providing improved visualizations of anatomical complexities relative to 2D physical models and 3D digital models [26]. A more involved usage of 3D printed models is for corroboration of various experimental models. Pack et al. studied whether a motion correction algorithm can improve the accuracy of dual-energy CT perfusion imaging in identifying and assessing coronary artery disease obstruction sites (by accounting for movement of the coronary arteries during diastole and systole that can limit diagnostic power of CT imaging) [27]. To this end, they implemented a 3D printed model of the left ventricle that allowed them to assess cardiac hemodynamics and test their motion correction algorithm, ultimately finding that their algorithm did help to correct for motion artifacts and identify coronary blockages with greater precision [27]. Wood et al. used a novel 3D printed phantom of brain vasculature to test the accuracy of various perfusion imaging modalities (e.g., digital subtraction angiography, CT, MRI) in assessing ischemia due to stroke [28]. They concluded that their 3D printed model aided in correcting for lack of standardization and inconsistencies between these various imaging modalities, thus improving overall accuracy of these perfusion measurement systems with the goal of treating ischemic stroke in a more timely fashion.

The applications of 3D printed constructs as experimental models extend beyond the studies discussed here. Thus far, these models have helped researchers to more accurately assess perfusion in vitro, thus facilitating medical discoveries that will help guide clinical decision-making. Future development of 3D bioprinted models would enable more accurate representations of native vascular and tissue architecture, and will only further improve in vitro assessments of perfusion.

FUTURE DIRECTIONS

In recent years, researchers have made significant progress bioprinting vascularized tissues that are increasingly representative of native tissue. However, there are still several challenges that must be overcome in future studies. Most bioprinting methods developed thus far have focused on fabricating homocellular constructs, which may be easier to study in vitro, but do not accurately represent the heterocellular tissue composition in vivo (e.g., blood vessels are made of three layers) [20]. Additionally, the hydrogels that comprise most bioinks inherently lack mechanical strength, requiring researchers to develop ways to impart mechanical integrity to their printed constructs (e.g., through extensive cross-linking) [11]. Furthermore, print resolution must also be improved. Current bioprinters are unable to achieve resolutions as high as 5 μm, and are therefore unable to replicate microvasculature of that scale (e.g., capillaries) [1]. It would also be of great value to develop a bioprinter that could deposit bioink at resolutions spanning several orders of magnitude, from nanoscale to microscale resolution, thus enabling printing of hierarchical vascular structures [21]. And although researchers have made vast improvements in the thicknesses of vascularized tissues that can be fabricated, they should also focus on improving speed of tissue fabrication (e.g., one method described by Kolesky et al. requires several days to create a construct on the scale of a human liver) [1,29].

The potential clinical applications of bioprinted vascularized tissues has been illustrated thus far primarily through proof-of-concept studies. Transplantation of bioprinted tissue has been successfully demonstrated in murine models, but it remains to be seen whether these successes can be reproduced in animal models of greater complexity, and eventually in human patients [20,24]. In order for bioprinted vascularized tissues to become more widely applied in providing and assessing perfusion as tissue replacements and experimental models, it is imperative for researchers to develop improved methods of tissue fabrication that address the challenges discussed above.

REFERENCES

[1] Paulsen SJ, Miller JS. Tissue vascularization through 3D printing: will technology bring us flow? Dev Dyn 2015;244(5):629—40. https://doi.org/10.1002/dvdy.24254.

[2] Zhu W, Qu X, Zhu J, et al. Direct 3D bioprinting of prevascularized tissue constructs with complex microarchitecture. Biomaterials 2017;124:106—15. https://doi.org/10.1016/j.biomaterials.2017.01.042.

[3] Miller JS, Stevens KR, Yang MT, et al. Rapid casting of patterned vascular networks for perfusable engineered three-dimensional tissues. Nat Mater 2012;11(9):768—74. https://doi.org/10.1038/nmat3357.

[4] Puchner SB, Liu T, Mayrhofer T, et al. High-risk plaque detected on coronary CT angiography predicts acute coronary syndromes independent of significant stenosis in acute chest pain: results from the ROMICAT-II trial. J Am Coll Cardiol 2014;64(7): 684—92. https://doi.org/10.1016/j.jacc.2014.05.039.

[5] Jia W, Gungor-Ozkerim PS, Zhang YS, et al. Direct 3D bioprinting of perfusable vascular constructs using a blend bioink. Biomaterials 2016;106:58—68. https://doi.org/10.1016/j.biomaterials.2016.07.038.

[6] Fischbach C, Mooney DJ. Polymers for pro- and anti-angiogenic therapy. Biomaterials 2007;28(12):2069—76. https://doi.org/10.1016/j.biomaterials.2006.12.029.

[7] Richardson TP, Peters MC, Ennett AB, Mooney DJ. Polymeric system for dual growth factor delivery. Nat Biotechnol 2001;19(11):1029—34. https://doi.org/10.1038/nbt1101-1029.

[8] Mooney DJ, Lee KY, Peters MC, Anderson KW. Controlled growth factor release from synthetic extracellular matrices. Nature 2000;408(6815):998—1000. https://doi.org/10.1038/35050141.

[9] Giannopoulos AA, Mitsouras D, Yoo S-J, Liu PP, Chatzizisis YS, Rybicki FJ. Applications of 3D printing in cardiovascular diseases. Nat Rev Cardiol 2016;13(12): 701—18. https://doi.org/10.1038/nrcardio.2016.170.

[10] Ameri K, Samurkashian R, Yeghiazarians Y. Three-dimensional bioprinting. Circulation 2017;135(14). http://circ.ahajournals.org/content/135/14/1281.full.

[11] Dababneh AB, Ozbolat IT. Bioprinting technology: a current state-of-the-art review. J Manuf Sci Eng 2014;136(6):61016. https://doi.org/10.1115/1.4028512.

[12] Lee JM, Sing SL, Tan EYS, Yeong WY. Bioprinting in cardiovascular tissue engineering: a review. Int J Bioprinting 2016;2(2):136—45. https://doi.org/10.18063/IJB.2016.02.006.

[13] Borovjagin AV, Ogle BM, Berry JL, Zhang J. From microscale devices to 3D printing. Circ Res 2017;120(1). http://circres.ahajournals.org/content/120/1/150.full.

[14] Choi NW, Cabodi M, Held B, Gleghorn JP, Bonassar LJ, Strook AD. Microfluidic scaffolds for tissue engineering. Nat Mater 2007;6(11):908—15. https://doi.org/10.1038/nmat2022.

[15] Golden AP, Tien J. Fabrication of microfluidic hydrogels using molded gelatin as a sacrificial element. Lab Chip 2007;7(6):720. https://doi.org/10.1039/b618409j.

[16] Bellan LM, Pearsall M, Cropek DM, Langer RA. 3D interconnected microchannel network formed in gelatin by sacrificial shellac microfibers. Adv Mater 2012;24(38): 5187–91. https://doi.org/10.1002/adma.201200810.

[17] Hoch E, Tovar GEM, Borchers K. Bioprinting of artificial blood vessels: current approaches towards a demanding goal. Eur J Cardio-Thoracic Surg 2014;46(5): 767–78. https://doi.org/10.1093/ejcts/ezu242.

[18] Kolesky DB, Homan KA, Skylar-Scott MA, Lewis JA. Three-dimensional bioprinting of thick vascularized tissues. Proc Natl Acad Sci 2016;113(12):3179–84. https://doi.org/10.1073/pnas.1521342113.

[19] Lee VK, Kim DY, Ngo H, et al. Creating perfused functional vascular channels using 3D bio-printing technology. Biomaterials 2014;35(28):8092–102. https://doi.org/10.1016/j.biomaterials.2014.05.083.

[20] Ozbolat IT, Peng W, Ozbolat V. Application areas of 3D bioprinting. Drug Discov Today 2016;21(8):1257–71. https://doi.org/10.1016/j.drudis.2016.04.006.

[21] Mosadegh B, Xiong G, Dunham S, Min JK. Current progress in 3D printing for cardiovascular tissue engineering. Biomed Mater 2015;10(3):34002. https://doi.org/10.1088/1748-6041/10/3/034002.

[22] Ball O, Nguyen B-NB, Placone JK, Fisher JP. 3D printed vascular networks enhance viability in high-volume perfusion bioreactor. Ann Biomed Eng 2016;44(12): 3435–45. https://doi.org/10.1007/s10439-016-1662-y.

[23] Sooppan R, Paulsen SJ, Han J, et al. *In vivo* anastomosis and perfusion of a three-dimensionally-printed construct containing microchannel networks. Tissue Eng Part C Methods 2016;22(1):1–7. https://doi.org/10.1089/ten.tec.2015.0239.

[24] Mirabella T, MacArthur JW, Cheng D, et al. 3D-printed vascular networks direct therapeutic angiogenesis in ischaemia. Nat Biomed Eng 2017;1(6):83. https://doi.org/10.1038/s41551-017-0083.

[25] Trachtenberg JE, Santoro M, Williams C, et al. Effects of shear stress gradients on Ewing sarcoma cells using 3D printed scaffolds and flow perfusion. ACS Biomater Sci Eng February 2017. https://doi.org/10.1021/acsbiomaterials.6b00641.

[26] Lauridsen H, Hansen K, Nørgård MØ, Wang T, Pedersen M. From tissue to silicon to plastic: three-dimensional printing in comparative anatomy and physiology. R Soc open Sci 2016;3(3):150643. https://doi.org/10.1098/rsos.150643.

[27] Pack JD, Yin Z, Xiong G, et al. Motion correction for improving the accuracy of dual-energy myocardial perfusion CT imaging. In: Gimi B, Krol A, editors. International society for optics and photonics; 2016. 97880Z. https://doi.org/10.1117/12.2216986.

[28] Wood RP, Khobragade P, Ying L, et al. Initial testing of a 3D printed perfusion phantom using digital subtraction angiography. Proc SPIE—the Int Soc Opt Eng 2015:9417. https://doi.org/10.1117/12.2081471.

[29] Kolesky DB, Truby RL, Gladman AS, Busbee TA, Homan KA, Lewis JA. 3D bioprinting of vascularized, heterogeneous cell-laden tissue constructs. Adv Mater 2014;26(19):3124–30. https://doi.org/10.1002/adma.201305506.

[30] Murphy SV, Atala A. 3D bioprinting of tissues and organs. Nat Biotechnol 2014;32(8): 773–85. https://doi.org/10.1038/nbt.2958.

Surgical Predictive Planning Using 3D Printing

12

Muath Bishawi[1,2], Sreekanth Vemulapalli[3]

Department of Biomedical Engineering, Pratt School of Engineering, Duke University, Durham, NC, United States[1]; Division of Cardiothoracic Surgery, Department of Surgery, Duke University, Durham, NC, United States[2]; Division of Cardiology, Department of Medicine, Duke University, Durham, NC, United States[3]

INTRODUCTION

The field of cardiovascular intervention and surgery has experienced a tremendous improvement in outcomes over the last few decades. While the reasons for these improvements are multifactorial, one important feature has been the development of well-validated risk models that help clinicians in understanding each patient's individual risk of a certain procedure or an operation. For example, the society of thoracic surgeons' (STS) morbidity and mortality score has been widely adopted and represents an important tool for patient selection, counseling, and care. This ability to better risk stratify patients preoperatively has led to a number of quality interventions that improved overall outcomes for this patient cohort [1].

Despite these improvements, however, a number of limitations remain apparent. First, these models are highly dependent on the variables that are actually captured and used for model generation, validation, and calibration. Second, even if all relevant variables are captured in a risk score, it simply provides population estimates rather than patient-specific estimates. Importantly, patients with variant/unusual anatomy or characteristics are poorly represented in the derivation and validation of these risk models. As a result, the population and patient-specific performance of these models in patients with variant/unusual anatomy or characteristics is unclear. Furthermore, diseases where the anatomy is more complex introduce an additional layer of complexity and variations in imaging and treatment choices, quality of the interventions, and overall outcomes.

The mitral valve is a good example of this concept. Complex mitral valve disease requires expertise in image acquisition and analysis to help with very accurate quantification of the defect, surgical expertise and skill to perform complex repairs, and overall innovation of the care-team to think creatively about the planned intervention. Today's modern complex mitral repairs are highly dependent on advancements in echocardiography to better capture and represent the affected lesions, allowing surgeons a better understanding of defects and 3D anatomy. These

3D Printing Applications in Cardiovascular Medicine. https://doi.org/10.1016/B978-0-12-803917-5.00012-2

227

complexities, on multiple levels of the patient treatment process, introduce a large level of variability in outcomes. Another example is surgical aortic valve replacement (SAVR). For instance, the STS-predicted risk of mortality score for SAVR suggests that high risk of mortality is at $\geq 8\%$, meaning that 92% of patients will not experience a death event directly from their procedure. So how can we translate such numbers into actual actions on the patient being evaluated for surgical intervention? It is in such a situation where tools that help in administration of personalized therapy, such as 3D printing, become imperative.

The introduction of 3D printing offers the ability to preoperatively assess patients in a personalized manner based on objective data related to their anatomy and how that might result in modifications or fine-tuning of the planned operation/interventional procedure. There has been a growth in the use of 3D printing for surgical or interventional planning [2—4]. While much effort has been focused on the uses of 3D printing in congenital heart disease (discussed in Chapter 5), there has been a growing body of literature for the use of this technology for structural and valvular disease. Subtle anatomical relationships and personalized operative/interventional plans can now be better elucidated, well in advance of the operation. Furthermore, with the expansion of possible materials available for 3D printing applications, there is also a growth in the use of 3D printing for surgical "practice" for trainees, and even for experienced individuals in situations with complex anatomy [5,6].

The evolution of 3D printing in the cardiovascular space and its use for routine clinical practice will likely mirror that of transthoracic echocardiography (TTE) [7]. This evolution is seen at its early stages in 3D printing with simple proof-of-concept cases and the use by early adopters; but similar to TTE, 3D printing will eventually develop to become a routine tool in the overall treatment algorithm for numerous cardiovascular ailments. Eventually, 3D printing will reach the level of evolution to have outcome-based evidence of its utility using well-defined and measured endpoints (e.g., quality of life, costs, length of stay) (Fig. 12.1).

Step 1: Accurate reproduction of anatomy – visualization of interventional planning.

Step 2: Reproduction of physiology/interaction with model – device sizing of choice.

Step 3: Demonstration of cost effectiveness and marginal value of pre-existing clinical variables.

Step 4: Specific device manufacture with selection of printer type, material choice best suited for desired application (training vs. implantation).

FIGURE 12.1

Predicted stages of development for 3D printing in adult cardiovascular disease.

Accordingly, this chapter will address: (1) cardiac imaging modalities for acquisition of 3D datasets for 3D printing, (2) current clinical applications of 3D printing in cardiothoracic surgery, and (3) future directions for development and improvement of 3D printing for clinical application.

IMAGE ACQUISITION MODALITIES

In order to accurately capture the anatomy and physiology required to create accurate 3D prints that can be used for surgical planning and training, appropriate image acquisition and processing techniques must be employed. Within the cardiovascular space, all four primary medical noninvasive imaging techniques (angiography/X-ray, ultrasound, computed tomography [CT], and magnetic resonance imaging [MRI]) are used.

Among these, the most commonly used cardiovascular imaging technique is transesophageal or transthoracic echocardiograms (TEE or TTE). With the advent of widely available clinical 3D TTE and 3D TEE systems in the 2000s, acquisition of 3D cardiac datasets using ultrasound, which could potentially be used for 3D printing, became a reality. Some benefits of using TTE or TEE for 3D image capture are the following: (1) lack of iodinated contrast requirement decreasing secondary side effects, (2) high temporal resolution, (3) definition of the tissue rather than the blood volume, and (4) lack of radiation exposure to the patient.

TTE/TEE have been used by some investigators to evaluate their utility for image acquisition for the purpose of generating 3D prints. In a study by Zhu et al., TTE was used to produce 3D constructs of the heart in 44 patients undergoing cardiac surgery [8]. The prints were compared to 3D echocardiography, as well as direct intraoperative findings. Of the 44 patients evaluated, 4 had mitral prolapse, 2 had partial endocardial cushion defects, 2 had atrial septal defects (ASDs), 1 had a ventricular septal defect (VSD), 2 had rheumatic mitral stenosis, and 1 with tetralogy of Fallot (TOF). The remaining 32 did not have any diagnosed structural heart defect. Overall, the 3D printed models generated from TTE were equal to or more accurate than 3D echocardiography for all the disease categories [8]. While the investigation provided promising evidence of the accuracy of the TTE generated prints, it was not clear if they had any added value over typical 3D echocardiography.

Though echocardiography is the most commonly used cardiac imaging technique, currently, the vast majority of clinical 3D printing is derived from CT. For the purposes of cardiac 3D printing, CT requires the addition of cardiac gating (4D cine CT) for optimal results. Advantages of gated cardiac CT over echocardiography for 3D printing include: (1) improved spatial resolution, unencumbered by imaging "windows," (2) rapid 3D dataset acquisition, (3) completely noninvasive nature (as compared to TEE), (4) improved image quality as compared to TTE/TEE, (5) reduced variability in image quality/image acquisition, and, (6) ease of postprocessing. Specifically, multiple free and paid image processing software platforms exist to transform tomographic images into formats required by 3D printers (*.stl), whereas few platforms exist which are capable of processing 3D Cartesian formats produced by current clinical echocardiography vendors.

Cardiac MRI can also be used to produce 3D datasets suitable for 3D printing, and its advantages over echocardiography largely mirror those of cardiac CT, with a few exceptions. Cardiac MRI has the advantage of not exposing patients to radiation, while additionally being able to generate 3D blood volume datasets with the use of renally safe contrast or even without contrast. Conversely, cardiac MRI requires specialized expertise in image acquisition, is time consuming, and may be contraindicated in certain patients with preexisting surgical implants.

CURRENT APPLICATIONS OF 3D PRINTING

Left Atrial Appendage Exclusion

The development of percutaneous devices for left atrial exclusion, especially for patients with atrial fibrillation (afib), has gained much interest recently. Surgical planning for such cases can be complex and relies heavily on accurate imaging of the location, size, shape, and unique characteristics of the ostium and body of the left atrial appendage (LAA). Additionally, for endovascular cases, planning and location of the trans-septal puncture is of prime importance. The position of the trans-septal puncture determines the distance to the LAA ostium and the angle at which the catheters will enter the ostium. Entrance angle is key in determining the possibilities for how the device can be placed (angle, depth, etc.). Additionally, entrance angle and distance are important in choosing among multiple available catheter curvatures (Fig. 12.2). Modeling these features prior to entering the operative arena may help in shortening the overall operative time (and associated risks/costs), and decreases the risk of poor deployments with leaks. It is therefore no surprise that 3D printing offers unique advantages for such a patient cohort [9−13]. This is especially true given the large variation in the size and shapes of the ostium and body of the LAA across the population (Fig. 12.3). Additional variations are also seen in the relationship between the LAA and the pulmonary artery, which is key in determining the eligibility for some closure devices (i.e., LARIAT device) (Fig. 12.4).

| Single Curve | Double Curve | Anterior Curve |

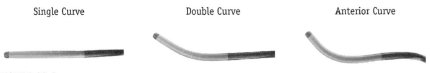

FIGURE 12.2

Differences in delivery catheter curvatures to help with occlusion device placement. 3D models can help with procedural planning and device selection leading to optimal placement during the actual procedure.

FIGURE 12.3

3D printed left atrium from the Duke Clinical Research Institute demonstrating the use of 3D printing for better visualization of the surrounding anatomy. *LA*, left atrium; *LAA*, left atrial appendage; *LUPV*, left upper pulmonary vein; *LV*, left ventricle.

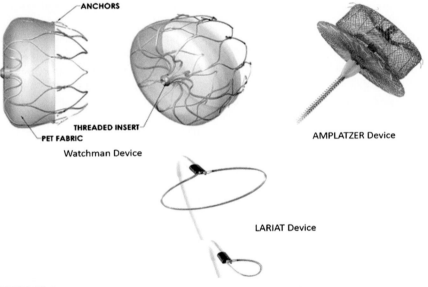

FIGURE 12.4

A number of different available occlusion devices. Preoperatively, 3D printed models can help in device selection leading to better left atrial appendage occlusion with minimal leakage on a patient to patient basis.

Reproduced with permission from Boston Scientific, St. Jude Medical and SentreHEART.

A case report examining the utility of 3D printing for LAA closure using the Wave Crest device (Coherex Medical, Inc., USA) and the Amplatzer Amulet device (St. Jude Medical, St. Paul, MN, USA) is illustrative of how 3D printing may be applied clinically in this field. The authors performed rehearsal of the operation using the devices with 3D printed LAA models generated using the respective patient images. Using such 3D printed models, the authors were able to select device sizing and examine the acquired compression and sealing of the LAA [9]. Importantly, oversizing of the device used was needed for one of the patients after the 3D model demonstrated poor coverage of the proximal LAA vestibule that was then corrected with a larger device. This led the authors to conclude that surgical planning using 3D printing offered substantial value over the available measurements acquired using angiography and TEE.

Another study by Hell et al. investigated if 3D printed LAA models led to better device sizing and selection [12]. They evaluated 22 patients who had afib but were at a high bleeding risk. Once the 3D LAA models were printed and the device (Watchman) deployed, they assessed device compression in a CT scan of the 3D model with the implanted device. The authors reported discrepancy between TEE and CT measurements with mean LAA ostium diameter on TEE being 22 ± 4 mm and 25 ± 3 mm for CT ($P = .014$) [12]. When compared with the final device size that was implanted, 21/22 (95%) of the 3D printed sizes were accurate, while only 12/22 (55%) of the 3D TEE sizes were accurate. TEE overall grossly underestimated the size of the devices used. Finally, the authors noted a strong correlation between device compression seen on the 3D LAA models and the actual clinical scenario ($16 \pm 3\%$ vs. $18 \pm 5\%$, r $= 0.622$, $P = .003$) respectively [12].

Lastly, Goitein et al. [11] generated 3D printed models, which were evaluated by three cardiologists in order to predict the size of the closure device to be used. This was done for 29 patients who eventually underwent closure with either the Amplatzer (n $= 12$ patients) or Watchman devices (n $= 17$ patients). In two patients, all three cardiologists predicted closure failure using the 3D model, which was actually true during the procedure. The final concordance correlation coefficient between the predicted size using the 3D models as compared to the final size used was 0.778 ($P = .001$) for the Amplatzer device and 0.315 ($P = .203$) for the Watchman device. Agreement between the three cardiologists was excellent regardless of device type (intra class correlation of 0.915 and 0.816) for the Amplatzer and Watchman devices, respectively [11].

Mitral Valve Repair/Replacement

Despite being in its infancy, percutaneous intervention (repair or replacement) of the mitral valve represents a promising area where 3D printing for operative planning or rehearsal might prove to be of significant value. While transcatheter aortic valve replacement (TAVR) has revolutionized our treatment algorithms for aortic valve disease, progress on percutaneous mitral valve replacement has been slow [14–16]. This has been mainly due to key differences between the aortic and mitral valves, including (1) the complex nature of the mitral valve anatomy, (2) mixed etiologies with less calcification compared to the aortic valve, (3) higher risk of valve migration, (4) relatively younger patients requiring mitral surgery compared to aortic, thereby increasing the need for long-term durability, (5) proximity and relationship to surrounding structures (i.e., occurrence of left ventricular outflow tract [LVOT] obstruction), and (6) the lack of a true rigid annulus to aid with anchoring and placement [17]. Additionally, many patients presenting with severe mitral regurgitation (MR) are essentially heart failure patients, with the majority of MR in the Western world being functional MR, as opposed to primary or so-called degenerative disease. Such patients have depressed ejection fractions (EFs), elevated pulmonary pressures, and require prompt treatment with low margins of error during such interventions.

Nowadays, there is an expanding toolkit of catheter-based treatment options. These include: the MitraClip device, transcatheter mitral valve replacement using an upside down TAVR valve, and early phase trials of dedicated mitral devices [18]. These early devices remain essentially surgical and are largely being deployed using a transapical route via a thoracotomy. A few case reports have demonstrated safety and some efficacy, with a major challenge being valve positioning and deployment [14]. Given such important limitations, 3D printing represents a unique opportunity for surgical planning and device practice.

A case report by Izzo et al. demonstrated the utility of 3D printing for catheter-based mitral valve replacement [19]. Two phantoms were generated (in 3 weeks) for a patient being evaluated for catheter-based mitral valve replacement. In one of the phantoms, a thin coating of a tantalum powder and superglue was applied to the print to mimic the radio attenuating properties of the patient's calcified valve. The other phantom print was left untouched for comparison purposes. The clinical team was able to practice valve deployment and positioning, which helped in the eventual successful implantation of the device.

While published data supporting the additive benefit of 3D printing in interventional or operative therapies of mitral valve disease is scarce, there is significant data supporting the benefit of 3D datasets in aiding surgical performance. In a single center study by Drake et al., institution of a 3D TEE perioperative imaging protocol was associated with a substantial increase in repair rates as compared to before institution of the protocol (46.6% vs. 77.6%) and as compared to the STS national average (54.9%) [20]. Furthermore, repairs were successful in 91.5% of isolated mitral operations after institution of the protocol as compared to the STS national average of 70.0%. Thus, while published data assessing the impact of 3D printing on measurable clinical outcomes in mitral valve disease is lacking, the study by Drake et al. suggests that 3D visualization does improve mitral valve outcomes on a case-by-case basis, and we envision that 3D printing may further improve upon advancements in 3D imaging.

The Tricuspid Valve

Similar to the mitral valve, a number of investigators have attempted to use 3D printing for the purpose of surgical planning and intervention in tricuspid valve disease [21]. Both surgical and percutaneous isolated tricuspid valve interventions remain rare [22]. For tricuspid valve interventions, the sizing and proper deployment of the valve is of utmost importance. As compared to the mitral valve, the tricuspid valve also has a saddle shape and constitutes leaflets, annulus, chordae, papillary muscles, and adjacent myocardium (atrial and ventricular). Unlike the mitral valve, however, the tricuspid valve is asymmetric, with three leaflets and a nonplanar, elliptical-shaped morphology, with the anteroseptal portion being superior to the posteroseptal one. Additionally, in secondary tricuspid regurgitation (the most common form), dilation occurs asymmetrically such that there is a greater increase in the anteroposterior diameter as compared to the mediolateral diameter. Given the anatomical complexity and the high comorbidity burden associated with tricuspid valve disease, 3D printing appears to have significant potential in aiding surgical or percutaneous therapies.

A number of groups have performed proof-of-concept experiments focused on attempting to generate 3D prints of the valve and its surrounding important structures using CT or TEE-based images [23]. In a case report by O'Neill et al., a patient who was considered at prohibitive surgical risk for traditional isolated tricuspid valve repair/replacement underwent transcatheter caval valve implantation. The group used 3D printing of the right atrium-inferior vena cava (RA-IVC) region to aid in interventional planning, as well as valve sizing [21]. Further work will be necessary to determine the additive value and clinical impact of 3D printing on tricuspid valve disease.

Cardiac Tumor Resection

3D printing can be utilized for more precise preoperative planning of complex tumor resection of the heart. In one report by Jabbari et al., 3D printing of two patients, who had large complex tumors involving the heart, was done preoperatively to better understand tumor size, extension, and involvement of other structures [2]. The first patient had a history of a left atrial osteosarcoma resection with mitral valve replacement, who had a recurrence 6 months later with four different lesions across different parts of the atrium and extending into the pulmonary veins. The second patient had a history of renal cell carcinoma who presented a year later with an encapsulated mass in the RA that extended into the IVC. Given the complexity of these patients in addition to the anticipated surgical resection and required reconstruction, both underwent 3D printing of their anatomy for surgical planning. The authors specifically commented on the ability to have depth perception with the 3D printed models and use of color-coded structures to help understand the margins of resection. This was especially important as a tumor extending too far into the pulmonary veins as in the first case, or too far into the IVC, as in the second case, would have required a much more complex operation.

Ventricular Septal Defects

Post-myocardial infarction VSDs represent a challenging clinical situation with significant morbidity and mortality. While surgical repair is standard therapy, such patients are at an exceedingly high surgical risk. Early surgical repair is of paramount importance as the septal defect continues to enlarge and remodel leading to surgical failure with delayed intervention. With advancement in percutaneous technologies, VSD closure devices could be used in the early phase post VSD diagnosis or in poor surgical candidates. Device size selection and positioning can be very challenging; however, presenting a unique situation where 3D printing for surgical planning may be beneficial. This is especially true given the large variability in degree of surrounding necrosis, shape of the VSD, and in some situations multiple isolated defects.

A case was reported by Lazkani et al. where a 3D printed model was constructed to help with percutaneous closure of an ischemic VSD [24]. A 57-year-old patient in cardiogenic shock had been transferred from an outside facility, and was diagnosed with muscular VSD 1 week post-coronary stenting for a myocardial infarction. Given that the patient was a poor surgical candidate, a percutaneous closure was attempted. The 3D printed model of the underlying anatomy was found to be helpful for full visualization of the defect, and a subsequent test was run prior to actual deployment of the device. In this case, the right-sided disc of the VSD occluder (20 mm) entangled with the right ventricular (RV) trabeculae during multiple attempts with the 3D model leading to incomplete coverage of the RV side of the defect. A different size was subsequently chosen leading to a more satisfactory outcome on the 3D model with more complete coverage. This change was then carried over to the clinical case leading to a successful closure of the defect [24].

Septal Myectomy for Hypertrophic Cardiomyopathy

Septal myectomy for hypertrophic cardiomyopathy is a less common procedure that is associated with a number of complications. The operative repair involves looking down the aortic valve after establishment of cardiopulmonary bypass, and performing nearly blinded cuts of the LVOT obstructing muscle. The procedure has a number of potential complications including a VSD (if too much tissue is removed) or a residual obstruction/gradient (if too little tissue is removed). Currently, the amount of tissue to remove is done by estimation and based on surgical experience. The impact of intraprocedural TEE is limited in this setting as cardiac standstill is instituted during myectomy, which precludes the use of Doppler to identify areas of turbulence and obstruction. As a result, assessment of surgical success normally occurs after the case is completed, which in most instances is too late to remove additional tissue if a residual gradient is seen. Additionally, multiple patterns of LVOT obstruction exist within hypertrophic cardiomyopathy, leading to unique outflow tract anatomies. Variant anatomy and difficult intraprocedural imaging combine to create an optimal situation for additive value through preprocedural 3D printing.

In a study by Hermsen et al., two patients with hypertrophic cardiomyopathy requiring septal myectomies had 3D printed models generated [3]. Simulation of septal myectomy was done on the 3D models prior to surgical resection. The authors reported congruence between the volumes of the resected tissue on the 3D printed model compared to the resected septal tissues from the patients. This study was unique since the printed models were used for surgical planning, as well as trainee simulation, for what is considered an operation that trainees will encounter infrequently during their training phase [3].

Transcatheter Aortic Valve Replacement

There has been a tremendous growth in the number of TAVRs performed [25–27]. As the valve delivery technology has been improving, there has been an expansion of this operation to lower risk patients. The most recent trials comparing TAVR to SAVR have demonstrated noninferior short and mid-term results [26]. One main limitation of TAVR remains the occurrence of paravalvular leaks (PVLs) [28,29]. Multiple studies have identified PVL as an important predictor of morbidity and even mortality [29]. The tools to predict PVL prior to valvular deployment are limited.

A number of groups have applied 3D printing to aortic valve replacement [30,31]. Ripley et al. attempted to test the feasibility [30] of generating 3D printed phantoms of patients' aortic valve complexes using cardiac CT images. In 16 patients, they were able to generate 3D printed models, fit them with printed valves, and then studied for valve positioning and possible PVL using a light transmission test [30]. Importantly, the 3D printed models had excellent agreement for annular measurements with those made with 2D data. PVL was correctly predicted in 6 of the 9 patients (67%) that eventually developed PVL after TAVR deployment. This study had a number of important limitations, however, including a lack of a correlation between the actual severity of PVL and the predicted size on the 3D prints. Furthermore, in two patients, the models had a false positive prediction (i.e., model showing possible PVL, while the clinical case did not experience PVL). This could have been a limitation of the 3D printed models, as often times, leaflet calcifications can play a role in "plugging in" possible PVLs. Some of these instances can also be due to limitations of the light transmission test. In our experience, the use of light transmission as a method to quantify PVL is limited by the aortic geometry.

In a subsequent study by Qian et al., pre-TAVR CT scans were used to generate aortic root 3D prints for 18 patients undergoing TAVR [31]. The 3D printed models were then used to deploy the CoreValve device and once full expansion was achieved, they were evaluated. The group developed an annular bulge index by assessing the annular strain on the 3D print generated by the valve. They found a strong correlation between the maximum annular bulge index and the development of PVL (and PVL severity) after the clinical valve deployment. This index was also able to predict the PVL location in 75% of patients [31]. Given the importance of PVL in determining outcomes after TAVR and the increasing commercial availability of multiple TAVR valve manufacturers and valve sizes, patient-specific choice of optimal valve manufacturer and valve size represents a present and future indication for cardiac 3D printing [29].

Adult Congenital Heart Disease Requiring Ventricular Assist Device Implantation

Individuals with uncorrected cardiac anomalies represent a challenging cohort of patients. A large portion of adult individuals with congenital anomalies eventually develop heart failure, with recent estimates showing nearly 1 in 4 by the age of

30 [32]. Many of these adult congenital patients, especially those with TOF, end up needing electrophysiology devices as well (i.e., implantable cardioverter defibrillators [ICDs]). Similarly, Fontan patients often need atrial fibrillation/atrial flutter ablations in the RA (since most of atrial arrhythmia arises from the RA rather than the left side). Any intervention on such patients is much more complex due to variant cardiac, systemic venous, and pulmonary arterial/venous anatomy, thus warranting additional tools to help clinicians navigate patient-specific anatomy.

A good example of this complexity is the use of ventricular assist devices (VADs). A growing portion of these adult congenital patients are undergoing VAD implantation, often a complex and high-risk operation. Accurate positioning of the VAD inflow cannula is difficult in such patients, especially in cases where the right ventricle is the failing systemic ventricle. The surgeon is faced with imprecise location of insertion and significant sources of obstruction to VAD inflow due to trabeculations and the presence of a moderator band. Farooqi et al. described the printing of 3D models in three patients who required VAD implantation. Such a step was reported to have significant implications regarding the accuracy of the insertion location and final positioning [33]. Given the growing adult population with congenital heart disease, the prevalence of end stage heart failure, and the shortage of organs available for cardiac transplantation, 3D printing to guide Left Ventricular Assist Device (LVAD) implantation is expected to become an area of significant interest and future research [34].

FUTURE DEVELOPMENTS IN 3D PRINTING FOR CARDIAC SURGERY

Real-Time Printing of Patient-Specific Implantable Cardiovascular Devices

The field of 3D printing in cardiovascular disease and cardiothoracic surgery remains in its infancy and has been largely focused on surgical planning and practice. At present, cardiac and cardiothoracic applications of 3D printing largely represent alternative forms of noninvasive diagnostic imaging. However, the potential for growth and development in the field is tremendous. Theoretically, 3D printing can be extended from modeling of complex anatomy to other applications such as modeling of physiology, when combined with pumps to recapitulate hemodynamics [35]. Additionally, as materials available for 3D printing advance, accurate simulation of in-vivo properties (elasticity, compressibility, etc...) of relatively static tissues such as blood vessels and LAA tissue in the fibrillating atrium will become more easily achievable. Advancements in 3D printing materials will allow for easier recapitulation of accurate physiology, and will also allow for use of 3D printed models to accurately predict endovascular and surgical device implantation characteristics (sizing, compression, potential tissue perforation, etc.) preprocedurally.

Finally, further developments allowing for increased availability and affordability of 3D printers capable of printing in metals and composites may eventually allow for on-site fabrication of patient-specific implantable cardiovascular devices and physiologically accurate models. In other surgical domains, patient-specific

3D printed skulls and jaw prostheses have been surgically utilized [36–39]. In the cardiovascular arena, patient-specific implants can be created for complex structures (printing the structure as a whole), such as the aortic root with an aortic valve or a mitral ring, or an entire valve on its own. To this end, clinicians working with engineers and biotechnology startups must come up with reproducible, easy-to-use methodologies to streamline image acquisition, processing, and production of high-quality 3D printed models and prostheses. Recognizing the potential future impact of 3D printed medical devices, the Food and Drug Administration recently released a guidance document for the development of such devices [40]. A proof-of-principle demonstration for the implantation of a patient-specific LAA occluder in a canine model was developed using 3D printed molds generated from CT scan [41]. This work paves the way for future disease states and devices fabricated using 3D printing.

Knowledge Gaps in 3D Printing in Cardiothoracic Disease

As advancements in 3D printing materials and 3D printer technology broaden the horizon in the cardiovascular space, data to support the impact of these applications on care outcomes will be necessary to prove efficacy and justify inclusion in clinical care. Currently, as we have demonstrated in this chapter, the majority of examples of 3D printing in the cardiac and cardiothoracic literature suggest that 3D printing is helpful for confirmation of underlying anatomy and planning of interventions. Yet, few of these studies report measurable improvements, such as increased diagnostic accuracy, reduction in procedural times, increased procedural success or decreased complications. Studies that do report these metrics are generally confined to case reports or small case series without historical or concurrent controls – making evaluation of the impact of 3D printing on outcomes difficult. Other, more difficult-to-achieve outcomes of interest for studies evaluating 3D printing may include impact on mortality, cardiovascular mortality, heart failure hospitalization, disease appropriate surrogates (for example, residual MR in surgical mitral valve repair or PVL in TAVR), or costs of care (Fig. 12.5).

Financial Viability of 3D Printing and Integration Into Routine Cardiothoracic Practice

To ensure the financial viability of integrating 3D printing into routine cardiac and cardiothoracic care, data regarding the impact of 3D printing on clinical outcomes and healthcare cost will be needed. At present, most centers performing 3D printing for use in clinical practice do so for selected cases, using investigator discretionary funds, research grant-related funds, or through medical center subsidies. Unfortunately, none of these funding sources is sufficient to allow for sustained, routine integration of 3D printing technologies into clinical practice. In the current era of cost consciousness and focus on quality of care and clinical outcomes, data supporting the impact of 3D printing on measurable clinical outcomes will need to improve to justify reimbursement by insurers and/or major investments by healthcare systems. Alternatively, 3D printing by clinicians and medical centers will remain

Proposed outcomes for assessing the impact of 3D printing in cardiac and cardiothoracic practice

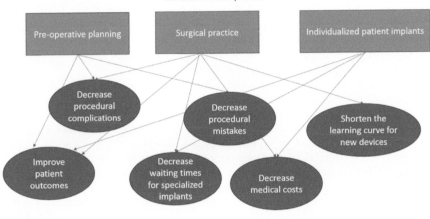

FIGURE 12.5

Predicted added benefits of 3D printing in cardiovascular diseases.

a niche technology with the major benefit being reaped through lower production costs of for-profit patient-specific devices and prostheses manufacturers.

CONCLUSIONS

While there are currently multiple applications of 3D printing in the cardiovascular space, the field remains in its infancy with the majority of contemporary applications being limited to the reproduction of complex anatomy in order to aid with patient-specific preprocedural planning. Current applications span the breadth of cardiac/cardiothoracic interventions from LAA occlusion, valvular heart disease, septal myectomy in hypertrophic cardiomyopathy to VAD placement in adult congenital heart disease. These successes indeed represent a first step towards actualization of personalized medicine in cardiovascular medicine and surgery. However, future improvements in the availability and affordability of 3D printers capable of printing in multiple materials, suited for the reproduction of tissue characteristics, will allow for better reproduction of physiology and greater utility in the selection and production of cardiovascular devices. To fully realize the promise of 3D printing, these advancements will require studies demonstrating the impact of 3D printing on patient outcomes.

REFERENCES

[1] Winkley Shroyer AL, et al. The society of thoracic surgeons adult cardiac surgery database: the driving force for improvement in cardiac surgery. Semin Thorac Cardiovasc Surg 2015;27(2):144—51.

[2] Al Jabbari O, et al. Use of three-dimensional models to assist in the resection of malignant cardiac tumors. J Card Surg 2016;31(9):581−3.

[3] Hermsen JL, et al. Scan, plan, print, practice, perform: development and use of a patient-specific 3-dimensional printed model in adult cardiac surgery. J Thorac Cardiovasc Surg 2017;153(1):132−40.

[4] Son KH, et al. Surgical planning by 3D printing for primary cardiac schwannoma resection. Yonsei Med J 2015;56(6):1735−7.

[5] Valverde I. Three-dimensional printed cardiac models: applications in the field of medical education, cardiovascular surgery, and structural heart interventions. Rev Esp Cardiol (Engl Ed) 2017;70(4):282−91.

[6] Estevez ME, Lindgren KA, Bergethon PR. A novel three-dimensional tool for teaching human neuroanatomy. Anat Sci Educ 2010;3(6):309−17.

[7] Fordyce CB, Douglas PS. Outcomes-based CV imaging research endpoints and trial design: from pixels to patient satisfaction. JACC Cardiovasc Imag 2017;10(3):253−63.

[8] Zhu Y, et al. Preliminary study of the application of transthoracic echocardiography-guided three-dimensional printing for the assessment of structural heart disease. Echocardiography 2017;34(12):1903−8.

[9] Pellegrino PL, et al. Left atrial appendage closure guided by 3D printed cardiac reconstruction: emerging directions and future trends. J Cardiovasc Electrophysiol 2016;27(6):768−71.

[10] Otton JM, et al. Left atrial appendage closure guided by personalized 3d-printed cardiac reconstruction. JACC Cardiovasc Interv 2015;8(7):1004−6.

[11] Goitein O, et al. Printed MDCT 3D models for prediction of left atrial appendage (LAA) occluder device size - a feasibility study. EuroIntervention October 13, 2017;13(9): e1076−9. https://doi.org/10.4244/EIJ-D-16-00921.

[12] Hell MM, et al. 3D printing for sizing left atrial appendage closure device: head-to-head comparison with computed tomography and transesophageal echocardiography. EuroIntervention 2017;13(10):1234−41.

[13] Wang DD, et al. Application of 3-dimensional computed tomographic image guidance to WATCHMAN implantation and impact on early operator learning curve: single-center experience. JACC Cardiovasc Interv 2016;9(22):2329−40.

[14] Cheung A, et al. Short-term results of transapical transcatheter mitral valve implantation for mitral regurgitation. J Am Coll Cardiol 2014;64(17):1814−9.

[15] Cheung A, et al. 5-year experience with transcatheter transapical mitral valve-in-valve implantation for bioprosthetic valve dysfunction. J Am Coll Cardiol 2013;61(17): 1759−66.

[16] Banai S, et al. Transapical mitral implantation of the Tiara bioprosthesis: pre-clinical results. JACC Cardiovasc Interv 2014;7(2):154−62.

[17] Anyanwu AC, Adams DH. Transcatheter mitral valve replacement: the next revolution? J Am Coll Cardiol 2014;64(17):1820−4.

[18] Bapat V, et al. Early experience with new transcatheter mitral valve replacement. J Am Coll Cardiol 2018;71(1):12−21.

[19] Izzo RL, et al. 3D printed cardiac phantom for procedural planning of a transcatheter native mitral valve replacement. Proc SPIE-Int Soc Opt Eng 2016:9789.

[20] Drake DH, et al. Echo-guided mitral repair. Circ Cardiovasc Imaging 2014;7(1): 132−41.

[21] O'Neill B, et al. Transcatheter caval valve implantation using multimodality imaging: roles of TEE, CT, and 3D printing. JACC Cardiovasc Imag 2015;8(2):221−5.

[22] Zack CJ, et al. National trends and outcomes in isolated tricuspid valve surgery. J Am Coll Cardiol 2017;70(24):2953−60.

[23] Muraru D, et al. 3D printing of normal and pathologic tricuspid valves from transthoracic 3D echocardiography data sets. Eur Heart J Cardiovasc Imaging 2017; 18(7):802−8.

[24] Lazkani M, et al. Postinfarct VSD management using 3D computer printing assisted percutaneous closure. Indian Heart J 2015;67(6):581−5.

[25] Leon MB, et al. Transcatheter or surgical aortic-valve replacement in intermediate-risk patients. N Engl J Med 2016;374(17):1609−20.

[26] Kodali SK, et al. Two-year outcomes after transcatheter or surgical aortic-valve replacement. N Engl J Med 2012;366(18):1686−95.

[27] Makkar RR, et al. Transcatheter aortic-valve replacement for inoperable severe aortic stenosis. N Engl J Med 2012;366(18):1696−704.

[28] Reidy C, et al. Challenges after the first decade of transcatheter aortic valve replacement: focus on vascular complications, stroke, and paravalvular leak. J Cardiothorac Vasc Anesth 2013;27(1):184−9.

[29] Genereux P, et al. Paravalvular leak after transcatheter aortic valve replacement: the new Achilles' heel? A comprehensive review of the literature. J Am Coll Cardiol 2013;61(11):1125−36.

[30] Ripley B, et al. 3D printing based on cardiac CT assists anatomic visualization prior to transcatheter aortic valve replacement. J Cardiovasc Comput Tomogr 2016;10(1): 28−36.

[31] Qian Z, et al. Quantitative prediction of paravalvular leak in transcatheter aortic valve replacement based on tissue-mimicking 3D printing. JACC Cardiovasc Imag 2017; 10(7):719−31.

[32] Faccini A, et al. Heart failure in grown-up congenital heart disease. Minerva Cardioangiol 2018.

[33] Farooqi KM, et al. 3D printing to guide ventricular assist device placement in adults with congenital heart disease and heart failure. JACC Heart Fail 2016;4(4):301−11.

[34] Warnes CA, et al. Task force 1: the changing profile of congenital heart disease in adult life. J Am Coll Cardiol 2001;37(5):1170−5.

[35] Kolli KK, et al. Effect of varying hemodynamic and vascular conditions on fractional flow reserve: an in vitro study. J Am Heart Assoc 2016;5(7).

[36] Shao H, et al. Custom repair of mandibular bone defects with 3D printed bioceramic scaffolds. J Dent Res 2018;97(1):68−76.

[37] Roskies MG, et al. Three-dimensionally printed polyetherketoneketone scaffolds with mesenchymal stem cells for the reconstruction of critical-sized mandibular defects. Laryngoscope 2017;127(11):E392−8.

[38] Gao F, et al. Individualized 3D printed model-assisted posterior screw fixation for the treatment of craniovertebral junction abnormality: a retrospective study. J Neurosurg Spine 2017;27(1):29−34.

[39] Nguyen Y, et al. Modifications to a 3D-printed temporal bone model for augmented stapes fixation surgery teaching. Eur Arch Oto-Rhino-Laryngol 2017;274(7):2733−9.

[40] FDA. Statement by FDA Commissioner Scott Gottlieb, M.D., on FDA ushering in new era of 3D printing of medical products; provides guidance to manufacturers of medical devices. 2017.

[41] Robinson SS, et al. Patient-specific design of a soft occluder for the left atrial appendage. Nat Biomed Eng 2018;2:8−16.

The Future of 3D Printing in Cardiovascular Disease

13

Mohamed B. Elshazly[1], Michael Hoosien[2]

Division of Cardiology, Department of Medicine, Weill Cornell Medicine-Qatar, Education City, Doha, Qatar[1]; Department of Cardiovascular Medicine, Heart and Vascular Institute, Cleveland Clinic, Cleveland, OH, United States[2]

3D PRINTING AS A TOOL FOR EDUCATION AND SIMULATION IN CARDIOVASCULAR MEDICINE AND SURGERY

Medical education has evolved considerably throughout the course of the 20th century, with increasing emphasis on patient safety and minimizing adverse outcomes. Cardiovascular medicine has similarly seen tremendous change, especially in the invasive subspecialties such as structural interventional cardiology and cardiac electrophysiology. Today, simulation and procedural planning are both regarded as important components to improve both procedural success and patient safety, and 3D printing now provides useful tools to improve both of these aspects [1–4].

In addition to providing excellent anatomic detail—and thus structural relationships that mimic what students or trainees would see in a live patient or a cadaveric specimen—3D printing makes it possible to create models that are representative of patient-specific anatomy. This means that 3D printing models could be created to better comprehend complex cardiac anatomy or variants, such as congenital heart disease, anomalous coronary artery takeoffs, variable left atrial appendage (LAA) shapes, etc. In addition, 3D printing could be used to demonstrate patient cardiac anatomy before following a procedure, and would allow students to better understand the utility of various interventions, such as atrial septal defect (ASD) closure, LAA occlusion, or correction of complex congenital heart disease. 3D printed models could be critical for efficiently training interventionalists or surgeons to perform complex procedures on rare diseases, to which trainees get minimal exposure during training.

3D printed models could also be partitioned to create different cross-sectional profiles, which would allow trainees to visualize endocardial structures from various orientations. This would be especially useful for individuals training in structural interventional cardiology or electrophysiology, as it is critical for operators in these fields to understand how cardiac structures are oriented in various orthogonal imaging planes (such as the left anterior oblique, right anterior oblique, craniocaudal, and anteroposterior projections).

3D Printing Applications in Cardiovascular Medicine. https://doi.org/10.1016/B978-0-12-803917-5.00013-4

Overall, the cardiovascular field is becoming increasingly dependent on advanced imaging technologies, and 3D reconstructions can be readily created from commonly used imaging modalities such as computed tomography (CT), magnetic resonance imaging (MRI), and echocardiography. From a learning perspective, 3D printing makes it possible for trainees to better comprehend the relationship between cardiac imaging and in vivo anatomy, which is critical during catheter and surgical manipulation in the heart. 3D printed models could be readily assimilated into full simulation tools by coupling them to imaging tools, such as transesophageal echocardiography (TEE) or intracardiac echocardiography, and thus allowing trainees to visualize catheters inside replica hearts in conjunction with images they would be able to see during an actual procedure. Coupling 3D printed models of cardiovascular structures with imaging simulation would undoubtedly improve both trainee competence and patient safety. As the field of cardiovascular medicine and surgery transition further toward more complex interventions, this technology will play an increasingly important role in the training of future operators.

3D PRINTING CLINICAL APPLICATIONS IN THE FUTURE OF CARDIOVASCULAR DISEASE

Current applications of 3D printing in cardiovascular medicine and surgery have been discussed in previous chapters. In this chapter, we review how we believe such applications will develop in the future.

STRUCTURAL HEART DISEASE INTERVENTIONS

In daily clinical practice, clinicians have come to appreciate the complexity and interindividual variability of structural heart disease. In previous chapters, we have discussed potential applications of 3D printing technologies in this patient population. These include:

1. *Procedural planning* of transcutaneous valve replacements, percutaneous closure of atrial and ventricular septal defects (ASD, VSD), percutaneous closure of paravalvular leaks, and LAA occlusion procedures [2,3,5,6]. 3D printed patient-specific disease models would allow the operator to visualize optimal access strategies. They could plan for positioning of guide wires, access sheaths, and catheters, as well as optimize device sizes (Fig. 13.1). With future developments in 3D printing materials and technology, interventional cardiologists will have access to patient-specific 3D heart models that will allow them to optimally plan complex procedures for every unique patient.
2. *Creating patient-specific custom-made devices.* 3D printing of patient-specific devices is currently of tremendous value in dental and maxillofacial procedures [7]. In the upcoming era of cardiovascular interventions, advances in 3D

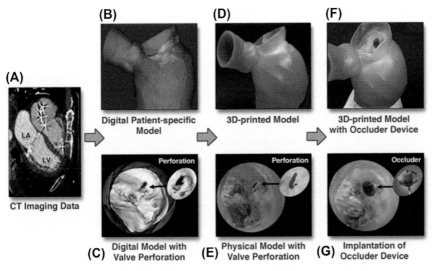

(A)

CT Imaging Data

(B) Digital Patient-specific Model

(D) 3D-printed Model

(F) 3D-printed Model with Occluder Device

(C) Digital Model with Valve Perforation

(E) Physical Model with Valve Perforation

(G) Implantation of Occluder Device

FIGURE 13.1

CT images (A) are used to create a digital model (B) of the mitral valve with a perforation (C). A multimaterial patient-specific 3D model (D) was printed to replicate the mitral valve geometry, regional calcium deposition, and pathology (E). 3D printed model with implanted occluder device (F). Zoomed image of leaflet perforation repaired using an occluder device (G). *LA*, left atrium; *LV*, left ventricle.

Reproduced with permission from Little SH, et al. 3D printed modeling for patient-specific mitral valve intervention: repair with a clip and a plug. JACC Cardiovasc Interv 2016;9:973–5.

printing technology will allow us to custom-build prosthetic valves, ASD and VSD closure devices, and LAA occlusion devices that are sized and shaped to each patient's unique anatomy [2−4]. Patient-specific 3D printed prostheses would decrease the frequency of postprocedural device failure and complications. For example, the emerging technology of transcatheter mitral valve replacement is facing difficulties related to the complexity and interindividual variability of the mitral annular shape, presence of calcifications, subannular structures, and left atrial diameter [8,9]. As such, creating custom-shaped 3D printed transcatheter prosthetic mitral valves would help in precise valve deployment and decrease the rate of procedural failure and complications such as paravalvular leaks. Similarly, LAA anatomy is complex and varies significantly between patients [2,4]. Therefore, using 3D printing to build LAA occlusion devices suited to unique patients can potentially decrease device leaks and deployment failure [2,10]. While 3D multimodality imaging and postprocessing are currently advanced enough to allow creation of custom-built devices, huge advances are needed in 3D printing modalities and materials so that patient-specific 3D printed devices can reach clinical practice and meet the requirements of regulatory agencies.

3. *Creating patient-specific functional flow models.* In daily clinical practice, cardiologists are faced with complex cases where assessment of valvular disease severity can be challenging due to dynamic blood flow and physiologic conditions in the cardiac chambers. This is particularly evident in certain conditions such as low flow low gradient aortic stenosis (AS), as well as aortic regurgitation. Creating patient-specific 3D models of aortic valve disease using multimodality 3D imaging and multimaterial 3D printing, coupled with a flow phantom, would provide the ability to emulate a spectrum of aortic valve dysfunction. A recent study generated 8 patient-specific multimaterial 3D models of severe degenerative AS and assessed the functional performance of each model under different in vitro flow conditions. Each model closely replicated aortic valve anatomy with reproduction of calcium deposition, cusp thickening, and valve orifice shape; the study showed that the valve orifice area of the 3D printed model varied with increasing flow [11]. Therefore, patient-specific functional models may provide a controlled and reproducible testing environment for quantifying flow in different disease states. Similar functional flow models can be applied to test the performance of new cardiac devices or novel diagnostic measures such as CT-derived fractional flow reserve [3]. In the future, these patient-specific 3D models could become a cornerstone of preclinical testing for cardiac devices. This may provide significant benefits over currently utilized animal models, preventing the loss of life, and providing models that more accurately replicate human anatomy and physiology.

ELECTROPHYSIOLOGY APPLICATIONS

Catheter ablation is a common method for treating cardiac arrhythmia. There has been significant progress in 3D electro-anatomical mapping and ablation of complex arrhythmias over the last two decades. 3D virtual reality electro-anatomical maps of arrhythmia substrates and real-time catheter visualization using mapping systems have tremendously advanced the field, allowing better visualization and understanding of complex arrhythmia. However, the complexity of cardiac anatomy and arrhythmia substrates both endocardially and epicardially continues to be challenging for many operators. Thus, preprocedural patient-specific 3D models can help better elucidate the arrhythmia substrate and better target focal and reentrant arrhythmias effectively [2]. For example, building patient-specific 3D models of the left atrium, pulmonary veins, and LAA can create models that would help in choosing the best approach, catheters, and ablation technique (e.g., endocardial vs. epicardial vs. both) suited to every individual's unique anatomy. Models that delineate scar tissue preprocedurally can be especially helpful. This type of model can also improve our ability to access arrhythmias originating from complex structures such as papillary muscles and aortic cusps. Another important application in electrophysiology is creating patient-specific 3D models of coronary sinus anatomy that can be used as a roadmap for guiding optimal placement of the coronary sinus lead in cardiac resynchronization therapy. Such approach can shorten procedural time and reduce radiation exposure and contrast dose. Finally, 3D bioprinting of heart models in the future can potentially help delineate the electrical conduction system of the heart, allowing operators to target arrhythmias adjacent to critical structures such as the AV node or even optimize location of His-bundle pacing.

CONGENITAL HEART DISEASE

3D visualization of complex congenital heart disease is one of the most difficult tasks in cardiology, even with the use of multimodality imaging. The last few decades have witnessed the development of complex palliative and correctional procedures, which have increased the number of patients surviving to adulthood. Hence, patient-specific 3D models of complex anatomy, in congenital pediatric and adult patients, can be crucial in presurgical planning, choosing an optimal surgical access site, and spatial visualization of complex defects [4,12]. Moreover, patient-specific information about the size and shape of major vessels and preexisting shunts can be important for planning of palliative or correctional surgery, heart transplant, and implantation of ventricular assist devices [13]. The future of treatment of congenital heart disease may also include 3D printing of patient-specific devices (e.g., homografts, shunts, or prosthetic valves) that are optimally sized to every patient's age and unique anatomy. Certainly, the future of 3D printing applications in congenital heart procedures seems very promising, given the highly unique nature of each patient's anatomy, which requires patient-specific surgical planning and custom-made devices.

CORONARY AND SYSTEMIC VASCULAR DISEASE

Current advances in 3D printing technologies have allowed for the printing of 3D models of systemic vasculature and coronary arteries. In terms of coronary artery applications, flexible, hollow, 3D models can be built to allow for assessment of coronary flow hemodynamics and enable visualization and tactile perception of postintervention anatomy. These coronary models can be used to evaluate the ex vivo efficacy of noninvasive flow quantification techniques (e.g., CT angiography and CT fractional flow reserve) in areas of coronary artery disease (3, 4). They can also help in preprocedural planning of complex coronary interventions such as coronary bifurcation stenting [4].

In terms of systemic vasculature applications, 3D printing has shown clinical benefit in certain applications and procedures. In patients with Marfan syndrome and aortic root aneurysm, a cardiovascular 3D printing technology known as personalized external aortic root support (PEARS) placement can be used in place of an aortic root replacement procedure. Here, a custom knit fabric mesh is fabricated around a patient-specific model of the individual's aortic root created by rapid prototyping, such that the mesh is perfectly shaped for that patient [14]. This patient-specific sleeve can then be surgically implanted to fit the patient's aortic valve and root morphology. Another application of 3D printing is planning of endovascular interventions. Patient-specific 3D printed models can allow for the assessment of optimal stent dimensions, fenestrations, and positioning in cases of aortic arch and descending aortic aneurysms, particularly in cases with complex neck and distal anatomy [15]. In the future, we will likely be able to build patient-tailored endovascular stents with carefully designed fenestrations that would decrease rates of endoleaks and other post-implantation complications.

CURRENT CHALLENGES AND FUTURE DIRECTIONS OF 3D PRINTING IN CARDIOVASCULAR DISEASE

While the future of 3D printing in cardiovascular disease is bright, this technology is still at its infancy, and we need to make significant progress on several levels in order to harness its full potential (Table 13.1).

INNOVATION IN IMAGE ACQUISITION AND POST-ACQUISITION PROCESSING

3D printing in the heart depends on accurate acquisition of 3D volumetric cardiovascular images to accurately depict complex patient-specific cardiac structural geometry ideally in both static and dynamic states. Most models are currently generated from CT and/or MRI. 3D transesophageal and transthoracic echocardiography and rotational angiography have also been used to a lesser extent [16]. Ideally, combining several imaging modalities would be necessary to create the most accurate 3D replica of the heart. For example, volumetric assessment of cardiac chambers can be best achieved using CT or MRI, based on their resolution and image area. Conversely, valve anatomy is best assessed using echocardiography, due to the temporal resolution of this method, which can capture rapidly moving structures such as leaflets and papillary muscles. Each imaging modality has its own pros and cons with regard to image resolution and artifact that have been discussed in previous chapters. Thus, complex and time-consuming image acquisition and software processing is required to combine high-resolution images from several modalities while trying to minimize artifact. Postprocessing of 3D images in preparation for 3D printing is another complicated time-consuming process that involves using computer-aided design (CAD) manipulations such as wrapping or smoothing, extruding tissues or trimming to reveal anatomical structures, and adding connectors such as cylinders or splints between separate anatomical structures of interest [4].

The diversity and complexity of multimodal 3D imaging acquisition and image postprocessing for 3D printing oblige us to develop guidelines and recommendations to simplify, standardize, and streamline this process. This can be achieved by using platforms that incorporate all the software modules, such as image segmentation and CAD modeling. Improving postprocessing will also require commercial software packages that combine ease of use with high-end automated or

Table 13.1 Current Challenges and Future Directions of 3D Printing in the Heart

	Current Challenges	Future Directions
3D imaging	• Difficulty combining information from several modalities (CT, MRI, TEE) • Need for high resolution 3D imaging • High costs • Lack of standardized imaging protocols and workflow	• Develop software that combines information from several modalities • Improve cost and complexity of noninvasive multimodality imaging • Standardize imaging workflow as data becomes available
Post-processing of 3D imaging	• Complex and time-consuming process of postprocessing and computer-aided design software • Requires combining several modalities • Lack of standardized protocols and workflows	• Innovation in postprocessing software technology to allow incorporating multi-modality 3D imaging • Simplifying, automating and improving costs of postprocessing workflow • Standardize postprocessing workflow process for each specific application of 3D printing
3D printing materials	• Complexity of creating tissue-like material that is deformable, elastic and mimics cardiac flow conditions • Need for multimaterial printing to mimic disease states (e.g., calcified valves) • Complexity and dynamic nature of cardiac physiology across age, gender, loading conditions, and disease states • Complexity of building cardiac devices with various components and alloys (e.g: LAA occlusion devices) • Cost	• Innovation in material discovery and engineering to develop affordable materials that mimic cardiac tissue • Developing multimaterial 3D printing technology that allows for creating countless possibilities of anatomical variants and disease states suited to every unique patient • Import 3D printing technology used to build patient-specific devices from other fields into the cardiovascular field • Developing 3D printing material guidelines and recommendations • Establishing 3D printing centers of excellence capable of synthesizing and innovating in material engineering • Encourage private sector investments and dedicate grant funding to this emerging technology
Applying 3D printing in clinical practice	• Lack of guidelines and recommendations • Lack of robust evidence about clinical effectiveness and cost • Scarcity of 3D printing centers of excellence	• Combing data from several case reports • Designing unconventional "N of 1" trials that test the clinical benefit of a patient-specific approach to 3D printing applications • Design cost analysis studies • Establishing guidelines and recommendations on how to design studies that evaluate its clinical efficacy • Encourage private sector and national grant funding to fund studies in this emerging technology

LAA, left atrial appendage; TEE, transesophageal echocardiography.

semi-automated segmentation modules and medical 3D printing-geared CAD capabilities [4]. In addition, even with all of these functionalities, the properties of the resulting model can still be greatly influenced by the postprocessing protocols used by the operator. As a result, standardizing these processes will be crucial for supporting research efforts and encouraging innovation from several biotech sectors in 3D printing technology.

INNOVATION IN 3D PRINTING MATERIALS

One of the most important aspects of developing accurate 3D printed models of the heart is the ability to replicate versatile cardiac tissue composition and physiologic properties. Unlike its use for simple anatomical teaching, using 3D printing to plan complex surgical procedures or custom-build devices requires high modeling accuracy of tissue physiologic properties and minimal modeling errors. For example, these models need to accurately replicate the potential effect of variable physiologic conditions and the consequences of prosthetic valve deployment on native cardiac tissue. Moreover, they need to demonstrate the effect of endovascular stenting on vascular and coronary anatomy after stenting. Replicating cardiac and vascular tissue mechanics requires that 3D printed models be fabricated from more flexible tissue-like materials. Fortunately, significant advances in material engineering have created increasingly complex material blends that can closely mimic the mechanical properties of some cardiac and vascular tissues [4]. One important example is the TangoPlus family of materials (Stratasys) that are used to create malleable models of the heart with a broad range of stiffness and compliance specifications [4,17].

Multimaterial 3D printing technology, such as the PolyJet technology, is used to build complex anatomical structures by combining multiple colors and materials simultaneously [2,4,18]. PolyJet machines can use a blend of materials ranging from very soft (TangoPlus) to hard (VeroPlus) materials to create high-resolution models with diverse properties (e.g., calcifications on the mitral valve apparatus). The diversity and deformability of 3D printing materials is particularly important for building functional flow models capable of replicating hemodynamic conditions [11].

Despite the significant advances in 3D printing materials, the currently available materials can only replicate the mechanical and physiologic properties of cardiovascular tissue only to a certain extent. Moreover, native cardiac tissue is diverse and each has complex and dynamic mechanical properties that change with age and physiologic state (e.g., heart failure and hypertension). This area of material exploration is relatively new, yet rapidly progressing. Research into creating artificial 3D printed materials with a broad range of malleability and physiologic performance must show continued progress. Moreover, it is essential that these materials are affordable enough to be incorporated into 3D printing technology. One of the biggest hurdles towards progress in 3D printing technology is the high cost of printing materials and 3D printers capable of multimaterial printing. Thus, significant research in creating novel tissue-like and affordable materials is of crucial importance to advancing the field forward.

In addition to innovation in nonbiologic 3D printing materials, innovation in bioprinting and molecular printing will have tremendous impact on advancing 3D printing technology. These techniques are used to build engineered biological tissue constructs with complex and hierarchical structures, mechanical and biological heterogeneity that can closely mimic patient-specific cardiac structures as discussed in previous chapters [2]. Novel bionic materials include cellular suspensions and hydrogels, decellularized matrix components and microcarriers offering surface areas for quick cell attachment and scaffold-free cell spheroids enabling generation of more mature cardiac tissue models with high resilience and definition. However, 3D bioprinting is still a budding field of research with mostly in vitro applications [19]. Additional work is needed to create bioprinted tissues that replicate the complexity of native tissue. Furthermore, one of the greatest challenges in the field of bioprinting is creating tissue constructs that can transport nutrients to maintain the long-term viability of the tissue. Molecular 3D printing can also be an important cornerstone of precision medicine. This new field can create molecular 3D printed models of a patient's diseased organ or tissues with specific molecular targets for developing molecule-based diagnosis and treatment strategies. For example, patient-specific molecular 3D printed models of heart failure may help test the effect of several medications or interventions on cardiac tissue remodeling in unique patients. Advances in this field require significant progress in 3D printing technology, in addition to precision medicine tools such as proteomics and genomics.

INNOVATION IN BRINGING 3D PRINTING TO CARDIOVASCULAR CLINICAL PRACTICE

Despite the exciting future of 3D printing in cardiovascular medicine, certain barriers continue to prevent its widespread adoption in clinical practice. These include high costs, complexity of workflow, limited evidence that suggests clinical value, and lack of awareness about how it can change clinical practice. The field needs to make dramatic strides in producing robust evidence that proves the effectiveness and cost-efficiency of this technology before it can move to clinical practice. Leaders and scholars in the field should begin to develop general guidelines and recommendations on how to advance the field forward with short-term and long-term goals. Short-term goals include efforts to standardize 3D printing workflow, choose optimal and affordable 3D printing strategies and materials for use in specific clinical applications, establish 3D printing centers of excellence fully equipped with state-of-the-art 3D printing labs, and fully trained physicians, researchers, engineers, and technologists who will carry the field forward [2–4]. Long-term goals include establishing a robust body of evidence that would allow us to develop guidelines and recommendations for worldwide adoption of this technology. This body of evidence would initially start by consolidating case reports that use different 3D printing techniques and materials, and then progress into clinical trials that prove efficacy and cost-effectiveness in daily clinical practice. Given the personalized nature of 3D printing, novel and unconventional clinical trials are needed such as "N of 1" trials that examine the clinical and cost benefit both at the patient as well as population level [20].

Establishing these short- and long-term goals, in addition to establishing 3D printing centers of excellence with dedicated 3D printing labs and personnel, will be crucial in catalyzing development in this field and bring 3D printing to daily clinical practice.

CONCLUSION

3D printing has the potential to make paradigm shifts in cardiovascular medicine that bring us steps closer to establishing our aspiration towards personalized and precision medicine. The cardiovascular medicine community should be ready to embrace a future where personalized 3D heart models will allow us to give patients visual 3D tours of their diseased heart, improve teaching of cardiovascular anatomy and physiology to medical students and trainees, plan surgical and interventional procedures in patients with complex anatomy, and custom-build devices that can be deployed in the heart and vasculature with minimal complications and failure rates. Combining the efforts of experts in the field and establishing 3D printing cardiovascular centers of excellence will be necessary to develop general guidelines and recommendations that bring 3D printing clinical applications to the cardiologists and cardiovascular surgeons of the future.

REFERENCES

[1] Gosai J, Purva M, Gunn J. Simulation in cardiology: state of the art. Eur Heart J 2015; 36:777—83.

[2] Bartel T, Rivard A, Jimenez A, Mestres CA, Müller S. Medical three-dimensional printing opens up new opportunities in cardiology and cardiac surgery. Eur Heart J February 16, 2017. https://doi.org/10.1093/eurheartj/ehx016 [Epub ahead of print].

[3] Vukicevic M, Mosadegh B, Min JK, Little SH. Cardiac 3D printing and its future directions. JACC Cardiovasc Imaging 2017;10:171—84.

[4] Giannopoulos AA, Mitsouras D, Yoo S-J, Liu PP, Chatzizisis YS, Rybicki FJ. Applications of 3D printing in cardiovascular diseases. Nat Rev Cardiol 2016;13(12):701—18.

[5] Little SH, Vukicevic M, Avenatti E, Ramchandani M, Barker CM. 3D printed modeling for patient-specific mitral valve intervention: repair with a clip and a plug. JACC Cardiovasc Interv 2016;9:973—5.

[6] Schmauss D, Haeberle S, Hagl C, Sodian R. Three-dimensional printing in cardiac surgery and interventional cardiology: a single-centre experience. Eur J Cardio Thorac Surg 2015;47:1044—52.

[7]   Chen J, Zhang Z, Chen X, Zhang C, Zhang G, Xu Z. Design and manufacture of customized dental implants by using reverse engineering and selective laser melting technology. J Prosthet Dent 2014;112:1088−1095.e1.

[8]   Kheradvar A, Groves EM, Simmons CA, et al. Emerging trends in heart valve engineering: Part III. Novel technologies for mitral valve repair and replacement. Ann Biomed Eng 2015;43:858−70.

[9]   Regueiro A, Granada JF, Dagenais F, Rodés-Cabau J. Transcatheter mitral valve replacement: insights from early clinical experience and future challenges. J Am Coll Cardiol 2017;69:2175−92.

[10] Brzeziński M, Bury K, Dąbrowski L, et al. The new 3D printed left atrial appendage closure with a novel holdfast device: a pre-clinical feasibility animal study. PLoS One 2016;11:e0154559.

[11] Maragiannis D, Jackson MS, Igo SR, et al. Replicating patient-specific severe aortic valve stenosis with functional 3D modeling. Circ Cardiovasc Imaging 2015;8:e003626.

[12] Sodian R, Weber S, Markert M, et al. Stereolithographic models for surgical planning in congenital heart surgery. Ann Thorac Surg 2007;83:1854−7.

[13] Farooqi KM, Saeed O, Zaidi A, et al. 3D printing to guide ventricular assist device placement in adults with congenital heart disease and heart failure. JACC Heart Fail 2016;4:301−11.

[14] Treasure T, Takkenberg JJM, Golesworthy T, et al. Personalised external aortic root support (PEARS) in Marfan syndrome: analysis of 1-9 year outcomes by intention-to-treat in a cohort of the first 30 consecutive patients to receive a novel tissue and valve-conserving procedure, compared with the published results of aortic root replacement. Heart 2014;100:969−75.

[15] Schmauss D, Juchem G, Weber S, Gerber N, Hagl C, Sodian R. Three-dimensional printing for perioperative planning of complex aortic arch surgery. Ann Thorac Surg 2014;97:2160−3.

[16] Mahmood F, Owais K, Taylor C, et al. Three-dimensional printing of mitral valve using echocardiographic data. JACC Cardiovasc Imaging 2015;8:227−9.

[17] Hermsen JL, Burke TM, Seslar SP, et al. Scan, plan, print, practice, perform: development and use of a patient-specific 3-dimensional printed model in adult cardiac surgery. J Thorac Cardiovasc Surg 2017;153:132−40.

[18] Ibrahim D, Broilo TL, Heitz C, et al. Dimensional error of selective laser sintering, three-dimensional printing and PolyJet models in the reproduction of mandibular anatomy. J Craniomaxillofac Surg 2009;37:167−73.

[19] Aljohani W, Ullah MW, Zhang X, Yang G. Bioprinting and its applications in tissue engineering and regenerative medicine. Int J Biol Macromol February 2018;107(Pt A): 261−75.

[20] Schork NJ. Personalized medicine: time for one-person trials. Nature 2015;520: 609−11.

Glossary

A

Actin A protein that forms (together with myosin) the contractile filaments of muscle cells, and is also involved in motion in other types of cells.

Adventitia The outermost layer of the wall of a blood vessel.

Agarose A substance that is the main constituent of agar and is used especially in gels for electrophoresis. It is a polysaccharide mainly containing galactose residues.

Agenesis The failure of an organ to develop during embryonic growth and development due to the absence of primordial tissue.

Algorithm A process or set of rules to be followed for calculations or other problem-solving operations, especially by a computer.

Anastomosis A connection made surgically between adjacent blood vessels, parts of the intestine, or other channels of the body, or the operation in which this is constructed.

Angiogenesis The development of new blood vessels.

Angiography Medical imaging technique used to visualize the inside, or lumen, of blood vessels and organs of the body, with particular interest in the arteries and/or heart chambers.

Aortic valve A valve in the human heart between the left ventricle and the aorta. It is one of the two semilunar valves of the heart, the other being the pulmonary valve.

Atherosclerosis A disease of the arteries characterized by the deposition of plaques of fatty material on their inner walls.

Anteroposterior Relating to or directed toward both front and back.

Annulus A ring-shaped object, structure, or region.

Atresia Absence or closure of a natural passage of the body.

Atrial fibrillation An irregular, often rapid heart rate that commonly occurs as a result of disruption of organized atrial activation.

Atrial septal defect A birth defect that causes a hole in the wall between the heart's upper chambers (atria).

Atrioventricular node Part of the heart's conduction system that connects the upper (atria) and lower (ventricle) chambers.

B

Biocompatible (especially of materials used in surgical implants) not harmful to living tissue.

Bioprinting The use of 3D printing technology with materials that incorporate viable living cells, for example, to produce tissue for reconstructive surgery.

Blalock-Taussig shunt Surgical procedure used to increase pulmonary blood flow for palliation in duct-dependent cyanotic heart defects like pulmonary atresia, which are common causes of blue baby syndrome.

Bundle of His Bundle of cardiac muscle fibers that conducts the electrical impulses that regulate the heartbeat, from the atrioventricular node to the septum between the ventricles and then to the left and right ventricles.

C

Capillaries The fine branching blood vessels that form a network between the arterioles and venules.

Carbohydrates A large group of organic compounds occurring in foods and living tissues and including sugars, starch, and cellulose. They contain hydrogen and oxygen in the same ratio as water (2:1) and typically can be broken down to release energy in the animal body.

Carbon nanodots Small carbon particles.

Cardiomyopathy An acquired or hereditary disease of the heart muscle.

Cardiopulmonary bypass A technique that temporarily takes over the function of the heart and lungs during surgery, maintaining the circulation of blood and the oxygen content of the patient's body.

Cardiogenic shock A life-threatening medical condition resulting from an inadequate circulation of blood due to primary failure of the ventricles of the heart to function effectively.

Cardiomyocytes Contracting cells of the heart.

Chondrocytes A cell that secretes the matrix of cartilage and becomes embedded in it.

Chordae Cord-like tendons that connect the papillary muscles to the tricuspid valve and the mitral valve in the heart.

Clot A thick mass of coagulated liquid, especially blood, or of material stuck together.

Coaptation The drawing together of the separated tissue in a wound, fracture, or valve.

Collagen The main structural protein found in skin and other connective tissues, widely used in purified form for cosmetic surgical treatments.

Comorbidity The presence of one or more chronic diseases or conditions in an individual.

Computed tomography Diagnostic imaging which makes use of computer-processed combinations of many X-ray measurements taken from different angles to produce cross-sectional (tomographic) images (virtual "slices") of specific areas of a scanned object, allowing the user to see inside the object without cutting.

Confluence An act or process of merging.

Congenital heart disease An abnormality in the heart that develops before birth.

D

Decellularization A process used in biomedical engineering to isolate the extracellular matrix (ECM) of a tissue from its inhabiting cells, leaving an ECM scaffold of the original tissue, which can be used in artificial organ and tissue regeneration.

Dextran A carbohydrate gum formed by the fermentation of sugars and consisting of polymers of glucose.

Differentiation Specialization of a cell type due to expression of certain genes.

Dodecahedrons A three-dimensional shape having 12 plane faces; in particular, a regular solid figure with 12 equal pentagonal faces.

Doppler ultrasound A test in which high frequency sound waves are reflected off blood circulating within the cardiovascular system. The returning sound waves (echoes) are picked up and turned into pictures showing blood flow through the arteries or the heart itself.

Double-chambered right ventricle (DCRV) An uncommon congenital anomaly in which anomalous muscle bands divide the right ventricle into two chambers; a proximal high-pressure and distal low-pressure chamber.

E

Echocardiography The use of ultrasound waves to investigate the action of the heart.

Electrophysiology The study of the electrical properties of biological cells and tissues.

Electrospinning Method for producing fibers from a polymeric solution by using an electric draw force.

Endocardial cushion defect A congenital defect where a portion of the walls separating all 4 chambers of the heart is poorly formed or absent. Also, the valves separating the upper and lower chambers of the heart have defects during formation.

Endothelial cells The cells lining the inner layer of a blood vessel.

Extracellular matrix Collection of molecules secreted by cells to provide structural and biochemical support.

F

Fibrin An insoluble protein formed from fibrinogen during the clotting of blood. It forms a fibrous mesh that impedes the flow of blood.

Fluorescent Emission of light by a substance that has absorbed light or other electromagnetic radiation.

Fontan procedure A palliative surgical procedure in which venous blood is diverted from the inferior vena cava (IVC) and superior vena cava (SVC) to the pulmonary arteries without passing through the morphologic right ventricle.

G

Gelatin A virtually colorless and tasteless water-soluble protein prepared from collagen and used in food preparation as the base for jellies, in photographic processes, and in glue.

H

Hemodynamics Fluid dynamics of blood flow.

Heterocellular Formed by different types of cells.

Homocellular Formed by the same type of cells.

Hyaluron A polysaccharide molecule which is one of the chief components of connective tissue, forming a gelatinous matrix that surrounds cells.

Hydrogel A gel in which the liquid component is water.

Hydrophilic Having a tendency to mix with, dissolve in, or be wetted by water.

Hypertrophic cardiomyopathy A condition in which the heart muscle becomes abnormally thick.

I

Intracardiac echocardiography An imaging technique designed for real-time image guidance and visualization of anatomical structures, performed with the ultrasound transducer positioned within the cardiovascular system.

Iodinated contrast A form of intravenous radiocontrast containing iodine, which enhances the visibility of vascular structures and organs during radiographic procedures.

L

Leaflets Cusps of heart valves.

Left ventricular outflow tract A portion of the left ventricle of the heart through which blood passes in order to enter the aorta.

M

Magnetic resonance imaging A form of medical imaging that measures the response of the atomic nuclei of body tissues to high-frequency radio waves when placed in a strong magnetic field, and that produces images of the internal organs.

Matrigel Trade name for a gelatinous hydrogel secreted by sarcoma cells.

Mediastinum A membranous partition between two body cavities or two parts of an organ, especially that between the lungs.

Mediolateral Relating to, extending along, or being a direction or axis from side to side or from medial to lateral.

Microfluidics Both the science which studies the behavior of fluids through micro-channels, and the technology of manufacturing microminiaturized devices containing chambers and tunnels through which fluids flow or are confined.

Mitral prolapse Improper closure of the valve between the heart's upper and lower left chambers.

Mitral valve Also known as the bicuspid valve or left atrioventricular valve; it is a valve with two flaps in the heart that lies between the left atrium and the left ventricle.

Mitral regurgitation A backflow of blood caused by failure of the heart's mitral valve to close tightly.

Morbidity A diseased state or symptom; ill health.

Mortality The state of being subject to death.

Myocardial infarction Ischemic cell damage due to lack of blood flow to the myocardium, also known as a heart attack.

Myosin A fibrous protein that forms (together with actin) the contractile filaments of muscle cells and is also involved in motion in other types of cells.

N

Necrotic Dead tissue.

O

Optogenetics A technique in neuroscience in which genes for light-sensitive proteins are introduced into specific types of brain cells in order to monitor and control their activity precisely using light signals.

Origami The Japanese art of folding paper into decorative shapes and figures.

Ostium An opening into a vessel or cavity of the body.

P

Papillary muscles Muscles located in the ventricles of the heart.

Paravalvular leaks A leak caused by a space between the patient's natural heart tissue and the valve replacement.

Patent ductus arteriosus A heart defect in which a shunt between the pulmonary artery and aorta persist after birth.

Patient-specific Personalized intervention based on assessment of the patient on an individual basis.

Percutaneous Made, done, or effected through the skin.

Photochemical Relating to or caused by the chemical action of light.

Photoluminescence Emission of light from any form of matter after absorption of a photon.

Photosensitive Having a chemical, electrical, or other response to light.

Photopolymers A light-sensitive polymeric material, especially one used in printing plates or microfilms.

Piezoelectric Electric charge that accumulates in certain materials under mechanical deformation.

Pluripotent (of an immature or stem cell) capable of giving rise to several different cell types.

Polyalanine Organic polymer formed by a polymerization of alanine residues.

Polydimethylsiloxane A widely used silicone compound.

Polymers A substance that has a molecular structure consisting chiefly or entirely of a large number of similar units bonded together, for example, many synthetic organic materials used as plastics and resins.

Polypropylene A synthetic resin that is a polymer of propylene, used especially for ropes, fabrics, and molded objects.

Polypyrrole Organic polymer formed by a polymerization of pyrrole.

Proliferation Rapid reproduction of a cell, part, or organism.

Prostheses An artificial device that replaces a missing body part, which may be lost through trauma, disease, or congenital conditions.

Pulmonary valve The semilunar valve of the heart that lies between the right ventricle and the pulmonary artery and has three cusps.

Purkinje fibers Part of the heart's conduction system that synchronizes contractions of the ventricles.

R

Radiopaque (of a substance) opaque to X-rays or similar radiation.

Resection The surgical removal of part of an organ or structure.

Rheumatic mitral stenosis Narrowing of the heart's mitral valve caused by an infection called rheumatic fever.

S

Septal myectomy A surgical procedure performed to reduce the muscle thickening that occurs in patients with hypertrophic cardiomyopathy.

Shunts A hole or a small passage which moves, or allows movement of, fluid from one part of the body to another.

Situs inversus dextrocardia Condition that is characterized by abnormal positioning of the heart and other internal organs.

Stem cells Undifferentiated cells of a multicellular organism that are capable of giving rise to indefinitely more cells of the same type, and from which certain other kinds of cells arise by differentiation.

Stereolithography A form of 3D printing technology used for creating models, prototypes, and production parts in a layer-by-layer fashion using photopolymerization.

T

Tendinae Colloquially known as the heart strings are cord-like tendons that connect the papillary muscles to the valves in the heart.

Tetralogy of Fallot A rare condition caused by a combination of four heart defects that are present at birth.

Thermoplastic Denoting substances (especially synthetic resins) that become plastic on heating and harden on cooling and are able to repeat these processes.

Trabeculations A structural part resembling a small beam or crossbar.

Transcatheter aortic valve replacement (TAVR) Minimally invasive surgical procedure which replaces the valve without removing the old, damaged valve.

Transesophageal echocardiography An alternative way to perform an echocardiogram. A specialized probe containing an ultrasound transducer at its tip is passed into the patient's esophagus.

Transglutaminase An enzyme that catalyzes the formation of an isopeptide bond between a free amine group.

Transthoracic echocardiography The most common type of echocardiogram, where the probe (or ultrasonic transducer) is placed on the chest or abdomen of the subject to get various views of the heart.

Tricuspid regurgitation Leakage of blood backwards through the tricuspid valve each time the right ventricle contracts.

Tricuspid valve One of the two main valves on the right side of the heart. Normally, the tricuspid valve has three flaps (leaflets) that open and close, allowing blood to flow from the right atrium to the right ventricle and preventing blood from flowing backward.

Trigones (of the heart) A thickened area of tissue between the aortic ring and the atrioventricular ring.

Turbulence Flow of fluid characterized by chaotic changes in pressure and flow velocity.

U

Ultrasound A type of imaging that uses high-frequency sound waves to look at organs and structures inside the body.

V

Vasculature The vascular system of a part of the body and its arrangement.

Vasculogensis The process of blood vessel formation occurring by a de novo production of endothelial cells.

Ventricular septal defect A heart defect due to an abnormal connection between the lower chambers of the heart (ventricles).

Nomenclature

°C	degree Celsius
c	concentration
CMC	critical micelle concentration; concentration of surfactants above which micelles form and all additional surfactants added to the system go to micelles
Ca^{2+}	calcium
E	elastic modulus
g/mL	concentration
GPa	gigapascal
kPa	kilopascal
mmHg	millimeter of mercury; 133.322387415 Pa
mPa s	millipascal-second
$\eta\nu$	viscosity
MPa	megapascal
MW	megawatt; 10^6 watts
pH	potential of hydrogen; a scale of acidity from 0 to 14
ΔR	contractile stress
T	temperature
T_g	glass transition temperature
w/v	fraction of weight of solute in the total volume of solution
ψ	swelling function, ratio of dry to swollen hydrogel weight
ε_{ult}	ultimate strain
μm	micrometer

Author Index

Subject Index